SOTHIS PRESS

Sothis Press aims to completely reimagine the public domain book with an emphasis on universal access to captivatingly designed print editions. We believe that technology will enable every single book ever published to be perpetually accessible in beautiful print form. Sothis is beauty in ubiquity.

www.sothis.press
1 Waterhouse Square, London, ECIN 2ST

Modern Milling Machines, Their Design, Construction, and Working

A Handbook for Practical Men
and Engineering Students

Joseph Gregory Horner

SOTHIS PRESS

ISBN 978-1-0365-0067-2

Modern Milling Machines, Their Design, Construction, and Working
is in the public domain.

This edition is published by
Sothis Press © 2023. The company
is registered in England and Wales.
The Company Number is 14900389.

MODERN MILLING MACHINES

THEIR DESIGN CONSTRUCTION, AND WORKING

A HANDBOOK

For Practical Men and Engineering Students

BY

JOSEPH G. HORNER, A.M.I.MECH.E.

AUTHOR OF "PATTERN MAKING," "HOISTING MACHINERY," "TOOLS FOR ENGINEERS AND WOODWORKERS," "ENGINEERS' TURNING," ETC. ETC.

WITH 269 ILLUSTRATIONS

LONDON

CROSBY LOCKWOOD AND SON

7 STATIONERS' HALL COURT, E.C.

1906

Printed at THE DARIEN PRESS, *Edinburgh.*

PREFACE.

THE present period is one of growing specialisation in Technical Literature, as in Manufacturing. The books which give a general treatment of a particular subject appeal less to practical men than to students and amateurs. No apology, therefore, is necessary for issuing a work which treats of a single department of machine-shop practice, and one of great and growing importance. Its scope is very broad, as a perusal of the contents of this work will show. Milling machines have become highly specialised, and the work of milling is now subdivided between different groups of hands just as that of turning is, ranging from very plain to very difficult work.

The Author has treated lightly those sections of the subject which offer no special difficulties, and has given considerable space to the manufacture of cutters and the work of the machines that call for the exercise of special skill. A number of typical methods of holding work, as well as some fixtures and jigs, are shown, but more as a general guide to the machine-attendant, than with any thought of their covering a field that is immense in its extent and variety. Some of the latest improved machines are illustrated with fully detailed drawings reduced from workshop prints, and special attention has been given to the vexed question of obtaining speeds and feeds.

The thanks of the Author are due to the firms who have kindly supplied drawings and photographs of machines and operations.

JOSEPH G. HORNER.

BATH, *November* 1905.

CHAPTER I.

THE LEADING ELEMENTS OF MILLING MACHINE DESIGN AND CONSTRUCTION.

CHAPTER II.

PLAIN AND UNIVERSAL MACHINES.

CHAPTER III.

ATTACHMENTS AND BRACINGS.

CHAPTER IV.

VERTICAL SPINDLE MACHINES.

CHAPTER V.

PLANO-MILLERS OR SLABBING MACHINES.

CHAPTER VI.

SPECIAL MACHINES.

CHAPTER VII.

CUTTERS.

CHAPTER VIII.

MILLING OPERATIONS.

CHAPTER IX.

INDEXING, SPIRAL WORK, AND WORM, SPUR, AND BEVEL GEARS, ETC.

CHAPTER X.

SPUR AND BEVEL GEARS.

CHAPTER XI.

FEEDS AND SPEEDS.

MILLING MACHINES AND THEIR WORK.

CHAPTER I.

THE LEADING ELEMENTS OF MILLING MACHINE DESIGN AND CONSTRUCTION.

The Development of the Milling Cutter—The Utilities of the Milling Machine—The First Machine—Early Improvements—The Influence of the Emery Grinder in its Development—Existing Machines Classified—The Lincoln Miller—The Characteristics of this Type—Modified Forms—The Work of the Lincoln Miller.

The Development of the Milling Cutter.—The rapid development of milling processes during the lifetime of the present generation is one of the most remarkable and interesting facts in the history of workshop practice. At one time these processes were regarded by engineers of the old school with disfavour, and the belief was very general that they could have but a limited application in the formation of only small surfaces of comparatively simple outlines. Neither would much improvement have been possible but for the fact that the development of emery-grinding machines has kept pace with the elaboration of the milling cutters. Milling, therefore, affords an illustration of ideas long latent, good in themselves, failing of translation into general practice in consequence of necessary conditions not having developed sufficiently.

There is an early, possibly one of the earliest milling cutters in existence in America, made by Vaucanson, a famous French mechanic, born 1709, died 1782. It is pierced with a hexagonal hole, and its profile is approximately that of the cutters for gear-wheel teeth. The pitching of the teeth is very fine, more like that of a saw than a modern milling cutter, and they are irregular.

Mr Bodmer, of Manchester, it is stated, had made a milling machine so early as 1824. But the general use of milling cutters in England dates only from about twenty-five years past—since emery wheels were introduced for grinding the faces of the teeth.

The late Mr Dixon stated that, as the result of careful inquiries, he found that the use of the emery wheel for sharpening milling cutters was due to a Mr George Hannay, of Ulverstone, a brother of Mr Hannay of the firm of Schneider & Hannay, the original founders of the Barrow Steel Works.

The difficulties of cutter formation having been got over, there remained another equally great, relating to the construction of the machines. The depth of cut which can be taken by a tool, conditions remaining the same, diminishes with increased breadth of cutting edge. Deep and broad cuts do not coexist. Roughing tools are narrow, the finishing tools are broad. This fact holds good in relation both to single-edged tools, and to milling cutters. One result is that the latter cannot take the deep cuts that the former are capable of doing. More than that, when the width of such cutters increases beyond an inch or two; the stresses are so severe, even with shallow cuts, that vibrations are set up which strain the machines, and detract much from the accuracy of the work. This is the key to the difficulties which have been experienced by the builders and users of milling machines.

The Utilities of the Milling Machine.—The study of the operations of milling machines involves the consideration of a type of rotating tools differing in all respects from those used in drilling and boring. The boring head with cutters has some resemblance to a mill. In fact, some mills with inserted teeth are constructed very similarly to the boring head. But there the resemblance ceases. Mills have a larger number of cutters, and their functions are quite different. Some will bore circular work. But that is not the chief function of milling cutters. Their function is that of universal tools, capable of operating on surfaces plane, curved, regular or irregular, straight or spiral. In short, I can think of no tooling in the machine shop, save that of drilling, which is not also done by means of circular milling cutters. The circular form is, of course, merely a convenient method of arranging a large number of single cutters at equal distances around the axis, the mill being in effect a multiplication of single-cutting tool edges or points, each of which operates in quick succession without any return stroke. The number of teeth and the resulting diameter of a mill are therefore not of a hard-and-fast character, these being details which are settled by practical considerations and convenience. Thus, it would not matter in the case of most jobs whether a mill were 3 or 5 inches diameter, the results in the shape imparted to the work would be the same. Considerations of cost, stability, and class of machine used will determine the choice of one size in preference to that of another.

No milling machine has yet been constructed suitable for all classes of work. There are universal machines, capable of performing all varieties of operations; but they lack the stability necessary for the heaviest work, and the range requisite for many operations. Compromises in milling machines have not been very successful: hence it follows that for nearly every special class of work a special machine is obtainable, and the nature of the jobs to be done should always determine the selection of a machine.

Very much milling is of such a character that the cutters become subsidiary to measurement; either a single cutter may be made to definite dimensions, or two or more cutters may be arranged in series or gangs, with or without provision for adjustment for wear. The value of arrangements of this kind over single-cutting tools is apparent. With the latter there is a lot of preliminary experimental setting and adjustment required, much of which work, so far as dimension adjustments are concerned, is saved by the use of milling cutters. The question of higher cost of cutters in the first place has to be considered, but as the practice of milling increases, the relative cost of these diminishes.

Milling machines are of less value in general engineering shops than in those which deal with specialities. As these machines received their first development in the manufacture of pistols, rifles, sewing machines, and articles of kindred character, so these shops and those of analogous character still afford the best illustrations of the practice of milling. In some manufactories of this kind there will be scores of milling machines performing every conceivable kind of operation on iron, steel, and brass. The wide extent, precision, and economy of the operations performed must, however, be studied in several shops in order to be fully appreciated. In these firms, and others of which these are types, milling is reduced to a system in which interchangeability of parts is secured at a ridiculously low cost per piece. And in proportion as shop systems approach more nearly to these does the value of milling increase. Growing specialisation and the increase in number of small fittings are the conditions most favourable to this class of work. Such, however, is the tendency of modern engineering practice, and therefore that practice is favourable to the increasing use and development of the milling machine. Planing, shaping, slotting, all labour under the disadvantages of having a non-cutting return stroke, besides being unsuitable for cutting surfaces which have very irregular contours. And when these last are cut, they cannot be done at once, but must be produced in detail with traverse feeding. In the milling machine, on the contrary, broad surfaces of irregular contours can be cut with mills having those contours, without any traverse feed.

Yet, though it is true that milling lends itself more readily to special work than to that of a general character, the assumption has been too often hastily made that it has little chance in the general shop. It certainly would be most injudicious to make a radical change in the methods of a general shop already equipped with single-cutting tool machines. But it is wise to introduce milling gradually, beginning with those classes of work for which the long experience of other firms has proved them suitable. In any general shop there are a lot of articles which without doubt can be treated more economically by milling than by any single-tool machine. An enumeration of such articles is hardly worth attempting, but in the work of nearly any firm it would be easy to select scores of pieces to which milling would be eminently superior to any operation done on planer, shaper, or slotter. This remark is applicable not only to plane surfaces, but to those in which profile cutters can complete at one traverse many jobs which must otherwise be done by more than one cutter, and more than one series of traverses, and often by setting on more than one machine. Illustrations of these kinds will appear in later chapters; for the present it is sufficient to note the fact. There are also machines which are capable of dealing with heavy work—using long fluted cutters, or gangs of cutters, or face cutters with inserted mills; and there are many bulky castings which can be tooled more rapidly thus than with the single tools of the planer and slotter. Machines of this class are not of recent origin, but they are being developed and brought more under the notice of engineers, their value becoming better recognised than they were a few years since. These facts afford an indication that the milling machine is destined yet to occupy as important a place in the general shop as the single-cutting tool machines have done.

Milling has not nearly reached its possible developments with us yet. The planer, shaper, and slotter still hold their own, with little rivalry in some shops, and surfaces are slowly tooled which could be done more expeditiously and often with equal accuracy under a good system of milling. Many machinists would be astonished at the large areas and the intricate sections which are milled in some shop practice, where milling is the rule, and planing and shaping are of secondary importance. The real era of the milling machine will not arrive with us perhaps until the pressure of closer competition develops its yet latent possibilities. It is from milling and grinding that the greatest developments of machinists' work are to be anticipated in the future.

The selection of milling machines to perform general, or special functions is a question for individual choice, and must depend mainly on the requirements of any given shop. One of the favourite types to-day—the Brown & Sharpe—also one of the oldest, has either a wide range of functions or is absolutely universal. A machine equal to it in value for general work is the vertical one, with slotting machine type of framing, and compound tables. Strictly specialised machines are confined mostly to shops which do a special class of work. In the future, these, like other classes of machines, may be expected to become more common.

There is no machine in the shop which requires more skilled attendance. Even the gear cutters do not need such constant attention as the universal milling machines, because the work of the former is more repetitive than that of the latter. The care of the various milling cutters themselves, the methods of lubrication, the setting of the work, the best arrangements of tooling in order to produce in all cases the most accurate results, in the most economical ways or by the means at disposal, the relations of depth of cut and feed, the most suitable treatment for the quality of material being cut, the best form of cutters to use for any given job—these points have all to be settled by an intelligent apprehension of the circumstances of each case, and not by the perfunctory attention of the mere machine minder. There is an immensely greater variety of jobs on the universal machine than on others, the lathe excepted, and the mind must be always on the alert if the capacities of the machine are to be utilised to the utmost. It is in truth a beautiful piece of mechanism, the uses and application of which are ever being extended in the machine shop; and the greater the intelligence brought to operate it the greater is its value. It is therefore unwise for employers to give such machines in sole charge of lads or unskilled labourers, who will bring the machine and its work into disrepute. They should be attended to by intelligent mechanics, or by men who have had a good deal of experience in their charge. Unskilled men or boys can attend to machines doing plain or repeat work, but then they should simply be attendants, working under the supervision of another who is responsible for the scheming of methods, setting of work, regulation of feeds, and so forth. Machines doing a variety of work like those of the universal type can only be operated economically, and their full capabilities realised by highly skilled attendance.

One result of milling is increased responsibility thrown on the tool maker. Good tool makers are always in demand. The accuracy of results depends on the making of the tool, and its setting. Especially is this the case in the built-up gang mills. After this the attendant has little responsibility. The man who minds a slabbing machine therefore has less on his shoulders than the operator of a universal machine. Although the work of the miller and the planer are much in rivalry, yet the planer hand has more of initiative than the man who has charge of a plano-miller, when the cutters are made and set by a tool maker. The planer hand has to manipulate his single-edged tools to produce various shapes, the attendant at the milling machine may mind three or four machines after they have been started. The planer tools have mostly to be made to produce many shapes, the milling cutter will only reproduce its own outlines.

The milling machine should diminish the work of the fitter. Whether it does so or not depends on the way in which the milling is done. For it does not follow by any means that the possession of these machines carries the certain result of good work. That depends on the way in which they are treated. Always regard must be had to the power of a machine. To attempt to drive it quicker, and to feed harder than its capacity will justify, can only result in chatter and untrue work, which must afterwards be corrected by the fitter. It is not the mere revolution marks left by the cutters, which occur in all work, but the departure from truth due to the springing of the arbor and cutter, which will make surfaces convex, or otherwise distorted from the profile that the cutters should impart. Unduly heavy feeding, therefore, is incompatible with true work. Further, the care of the cutters themselves influences results. Even when properly formed and fluted, the maintenance of the cutting edges by grinding is a vital matter, since dull edges will not cut so truly as keen ones, due to the spring of the arbor.

Another point in connection with milling is this—that the hard scale on forgings and castings, which does so little injury to the single-edge cutting tools, for the simple reason that they are able to penetrate beneath it at once, is ruinous to the delicate edges of the milling cutters. Hence in every shop where milling is adopted as a system it becomes a matter of economy to remove the hard skin by pickling the iron and steel work in dilute sulphuric acid, and the brass work in nitric acid.

The capacities of the milling machine are more limited in one sense than those of the average machines in the shop. The reason is that an arbor of great length is objectionable because of the vibration set up in it by heavy, or even moderate, cutting. The size of arbors is limited by the diameter of the holes in the smallest mills. To increase the diameter of arbor would preclude the employment of small cutters, and restrict very much the utility of the machine. It becomes therefore impossible to put work of very large dimensions on ordinary milling machines. Exceptions occur in the heavy slabbing tools which do not fulfil universal functions, and the ending or rotary machines, which are used only for facing large work laid on a table alongside the machine.

Milling has conduced to general machining of a higher degree of excellence, both in respect of accuracy and of finish, than was formerly economically possible. Very much work that is done well and cheaply by milling was formerly done laboriously and imperfectly by the single-tool machines, or by hand, or by grinding.

The single-tool machines are accurate, but there is no one of them which is capable of performing so great a variety of processes as the milling machine. Neither is it possible, without taking two or three cuts, the last one with a broad tool, to produce so smooth a finish as a milling cutter will produce with one cut. Hand labour is too costly to hold a place against modern conditions, neither is it adapted for repetitive work. The excellent results which are producible by milling have rendered inferior work, by whatever method done, unjustifiable and inadmissible, so that a shop well provided with milling machines is able to turn out work not only better, but often cheaper, than a rival shop working by the older methods. In another way milling has acted on machine-shop practice. It has favoured the cheap production of repetitive work of small dimensions, so that uniformity of dimensions and form, and interchangeability, unknown and impracticable not so many years ago, are now becoming the rule.

The First Milling Machine.—The late Mr E. G. Parkhurst, known best by wire-feed reputation, but great in all that concerns the manufacture of small arms, has given an account of what is probably the first milling machine ever constructed. He saw the machine, and gathered the facts relating to it in 1851 from a Mr Robert Johnson, who gave the year 1818 as that in which it was first set to work. This was at a gun factory in Middletown, Conn., on a site known then, as now, as Mill Hollow.

At that period the file was the principal tool used by the gun makers, and the new milling machine was used as a roughing tool to lessen the labour of rough filing, leaving the finishing to be done by hand. Its application was confined to plane faces, and did not include either curves or profiled forms. It was a hand-operated machine for several years before a self-acting feed was added.

The machine, shown in Fig. 1, comprises the following parts:—A bed plate A, a rough casting measuring about 24 in. by 18 in. by 2 in. thick. A headstock B, removed from an engine lathe, was bolted to the bed, and carried a spindle with a square-tapered socket to receive the cutter spindle. The cutter was a plain one, with filed teeth. It measured about 1½ inches in diameter, by 1-inch face. The cone pulleys were of wood, of about 2½-inch face, the largest being about 8 inches in diameter. The drawing shows the grooved pulley at the rear for feed, a later addition to the first design.

The work was traversed under the cutter by the hand crank C, turning a pinion *a*, engaging with a rack screwed on the under side of the table D. The edges of the table were vee'd to fit the vee notches cut in the plugs *b*, *b*, carried in drilled holes in four uprights E, E, bolted to the bed, and secured by and rendered adjustable for wear by set screws *c*, *c*. The work was gripped in the piece F, fitted with pinching screws, having pointed ends. If a piece of work required a second cut, it was packed up in F with paper or sheet metal.

This machine, in existence three years after Waterloo, though crude in the extreme, is a venerable relic worthy of illustration in a book that deals mainly with present-day construction. Spite

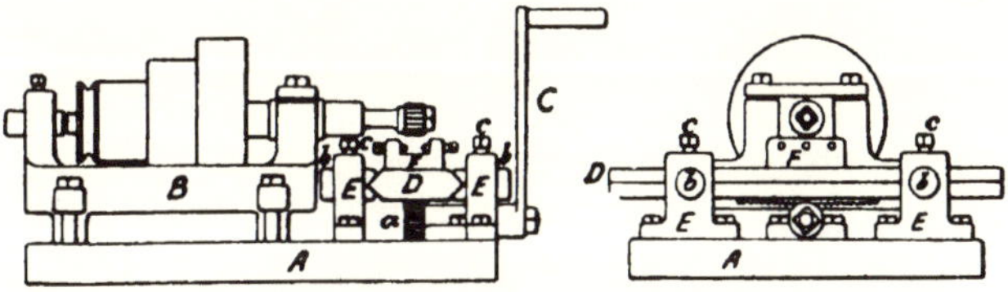

Fig. 1.—The First Milling Machine.

of its simplicity, the workmen are said not to have taken kindly to it at first, considering it an innovation on filing!

Early Improvements.—The first great improvement which was effected in the milling machine lay in the substitution of provision for vertical adjustments of the spindle bearings, for the fixed bearings of the first machines. This is said to have originated at the works of Mr Eli Whitney, of New Haven, Conn.

Little further progress was made until the early fifties, at which period there were a great many millers in existence, but mostly manufactured by firms for their own use. About this period they began to take rank as precision tools for duplicating pieces, previously to which they had retained their first function—that of roughing-down machines to the filers. Gang mills began to figure about this time, and an unsuccessful attempt was made to build a slabbing machine, somewhat on the lines of the planer. Soon after, 1850, the Howe machine was brought out, and then the Ames Manufacturing Co. of Chicopee, Mass., made numerous machines, one being a modified Howe, designed by Mr C. M'Farland. These machines much resembled in outline the Lincoln miller, designed by Mr F. A. Pratt in 1854, a type on which probably more machines have been made than of any other. Its distinguishing feature was the substitution of a screw feed in place of the pinion and rack feed hitherto applied. The quick return was effected by hand. In 1848 came the Root machine by Mr E. K. Root, superintendent of Col. Colt's armoury at Hartford. This was in the main built on the Lincoln and other earlier models, but the table was driven by a worm gearing into a spiral rack underneath the table. It embodied also a yielding clutch, with a hand lever for operating it, for the purpose of disengaging the driving cone pulley, so that the spindle could be stopped from the countershaft. There was also an automatic knock-off, which stopped the machine instantly, leaving the table standing at any desired part of its traverse. Some of these machines have remained in service for fifty years. Between 1861 and 1866, during a portion of which the American Civil War was raging, the hundreds of thousands of small arms turned out in government and private factories gave the opportunity for the improved milling machines, and they became firmly established as precision tools. At that period, too, sewing machines were booming, and so after 1863 nearly all the armoury-milling machines, which were being abandoned in consequence of the closing of the war, were purchased by the sewing-machine manufacturers.

Up to this period, therefore, forty-five years after the invention of the milling machine, there was but one type in use, the modern representative of which is the Lincoln miller. All these machines had horizontal spindles, and housings adjustable vertically. Most had tailstocks also, but the Root introduced an overhanging arm with movable centre, in place of the tailstocks. Back gears were fitted and compound tables. The legs in every case were of the **A** form, in lathe fashion. The pillar and knee machine had not appeared, nor the vertical spindle machines, nor any successful slabbing miller, nor rotary face mills with inserted teeth, nor any cutters of large dimensions.

The Influence of the Emery Grinder.—The milling machine would never have attained its present development but for the growth of the emery grinder. Its era dates from this period, notwithstanding that the milling cutter itself is more than a century old, and the milling machine was born about eighty years ago. In the older methods the temper of the cutters had to be drawn for sharpening, and rehardened, and this was inconsistent with accuracy, and with economy. It is to the emery wheel also that we owe the displacement of cast gears by cut ones. The old cutters were not accurate enough for such work, nor would they retain their shape unimpaired by grinding but for the form type of cutter.

Existing Machines Classified. — The types of machines, ranging from plain to universal, are as varied as the classes of work which are done by milling. These types also merge into one another, so that strict classification is not possible. The following will, I think, be a comprehensive and serviceable one :—

1. Lincoln millers.
2. Pillar and knee machines with horizontal spindles.
3. Vertical spindle machines and profiling machines.
4. Plano-millers or slabbing machines.
5. Special machines for gear cutting, &c.

The Lincoln Miller, Figs. 2 and 3.—This is one of the most common types of milling machine, dividing favour about equally with the pillar and knee types. As it was the first to be invented, it has retained a permanent place in favour notwithstanding the advent of new forms. Its great value lies in its stability and rigidity, the arbor being supported from the bed below, and the table being backed up by a solid bed resting on the ground, instead of by an overhanging knee. Its slight disadvantage is that the table and its work are partly behind the foot block or end support of the arbor, so that the workman cannot get round the job so readily as when an overhanging arm is used. But this does not interfere with the ordinary work put upon the miller, largely consisting of that which would otherwise go to the small planer, or to the shaper.

As the table has no provision for elevation in the Lincoln

machines, all movements of this kind are imparted to the spindle. The spindle bearings in the smaller machines slide in housings within the head and foot stocks, Fig. 2. In the larger ones they slide on the front upright faces of these heads, Fig. 3, each in-

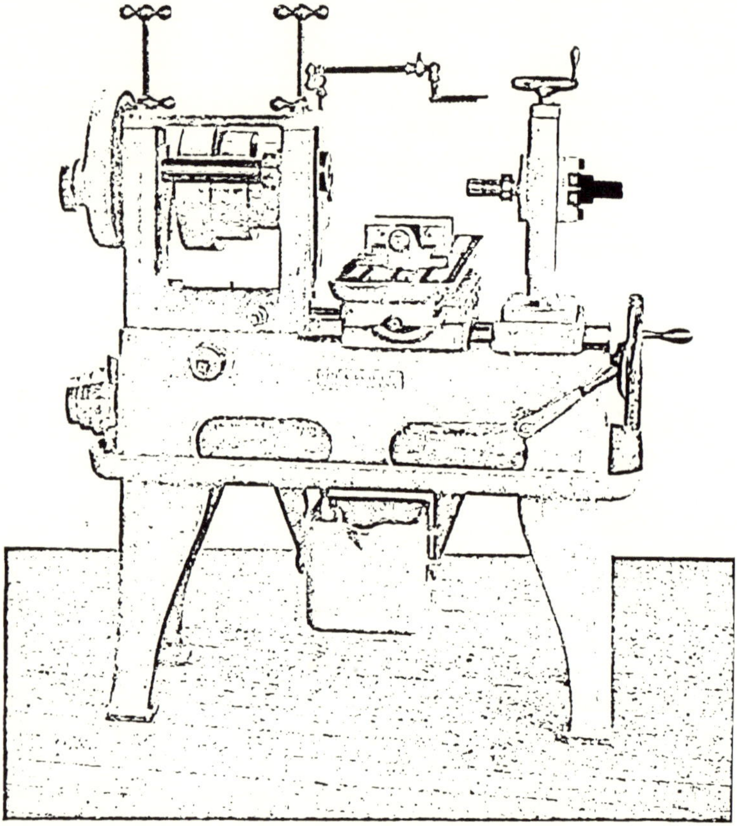

Fig. 2.—Lincoln Miller. (Pratt & Whitney Co.)

volving different operating mechanisms. Independent hand wheels and vertical screws are used, in some mitre gears operate the screws, as in the elevating mechanisms of planers. Frequently the bearings of the head and tail stocks are moved in unison by a single set of gear on the head, by the simple device of connecting the two spindle bearings with a stiff round bar, Fig. 3. This bar is made rigid enough to ensure the alignment of the spindle at all vertical positions. In some machines exact heights can be obtained by micrometric divisions of thousandths, or sixty-fourths of an inch. In one type of machine the dial boss revolves with the

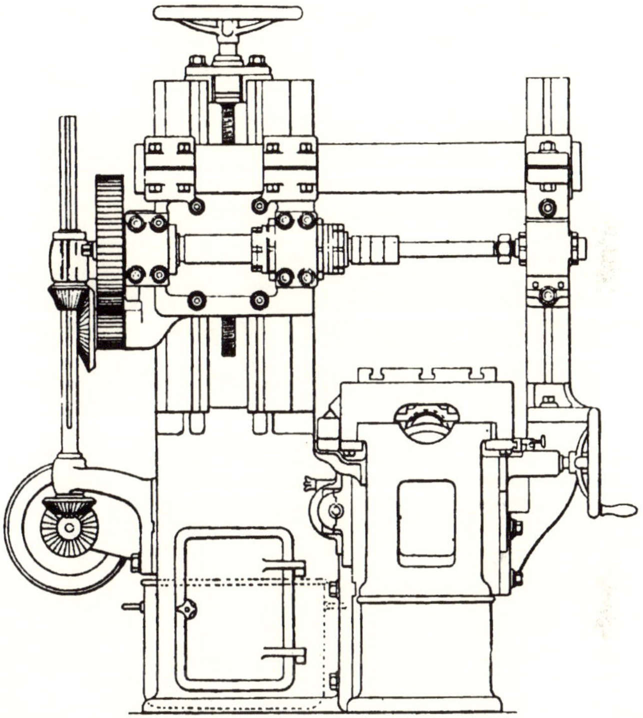

Fig. 3.—Lincoln Miller. (J. E. Reinecker.)

screw by a spring friction, but is adjustable by hand, the index finger remaining stationary. The mass of the spindle and bearings is counterbalanced in all but the smallest machines by weights suspended from chains passing over pulleys.

Generally the headstock is a fixture, sometimes cast with the bed, and the tailblock is adjustable to and from the head to a small amount, in order to permit of inserting cutters of different widths. The carriage of the table also has a lateral adjustment of three or four inches to accommodate the work slightly to the cutter after the latter has been fixed. In a few machines the headstock is also capable of some adjustment along the bed.

The forms of the beds vary. In the small machines the bed much resembles that of a lathe, Fig. 2, and is surrounded in modern machines with a deep waste-oil tray, and carried upon legs.

Fig. 4.—Modified Lincoln Miller. (Tangye Tool & Electric Co., Ltd.)

In larger machines the cabinet form of bed, Figs. 3 and 4, reaching to the ground is adopted.

The spindle drive is from stepped cones and through gears, provision being embodied to permit the spindle gears to engage at all heights. The feeds are derived from smaller stepped cones, the first of which is on the main cone shaft, the second on a shaft running alongside the bed, having a worm which drives a worm wheel, and thence through other gears the table feed screw. The feed is tripped, and the worm dropped automatically out of engagement at the end of a cut.

There are numerous modifications of this general type of machine. The principal differences are those due to increasing size. The machine begins to resemble the planer, Fig. 4, the head and foot stocks developing into two large housings bolted against the sides of a bed, which carries a table of greater proportional length than that of the ordinary Lincoln machine. These in strictness are not Lincoln machines, but they form an obvious link between these and the plano-millers. Allied to these are the double-headed milling machines, carrying horizontal spindles for face milling, and the machines with one head removable, so converting them into open-side machines.

With scarcely an exception the Lincoln machines are of plain type. There is an example of a universal machine, rendered so by the fitting of a swivel table, and a dividing head.

The Work of the Lincoln Miller.—The work put on this machine is generally of small dimensions, because the tables and their traverses are short in the standard machines. But within these limits there is a large volume which the machine will take. It has scored most in the work of the gunsmiths and sewing-machine makers, and then in engineers' work of a similar character. The greater volume done is horizontal milling, either plain or profile, the conditions being very favourable to the latter by reason of the good support given to the spindle and the table. Cuts as heavy as the teeth of the cutters will stand can be taken in consequence. Face milling is also done. In many cases the footstock is readily removable to permit of the insertion of wide pieces of work to be tooled thus. Small pieces of work are frequently arranged in series, up to the capacity of the machine.

CHAPTER II.

PLAIN AND UNIVERSAL MACHINES.

Pillar and Knee Machines with Horizontal Spindles—The Characteristics of this Type—Typical Machines Described—Difference between the Plain and Universal—Variable Details—Headstocks—Tables and Knees—Feeds—Whence Derived—Countershaft or Spindle—Numerous Examples—Micrometer Movements to Feed Spindles—Vernier Fitting to Tables—Differences between Plain Machines and Universals—Index Centres and Spiral Heads—Various Examples—Footstocks.

Pillar and Knee Machines with Horizontal Spindles.—The reason why so many, probably the majority of milling machines, are of the pillar and knee kind is traceable to the fact that the early milling was of a light character, for which machines of this type are eminently adapted. As this was among the earliest, so it is in its elements the simplest form of machine. Fig. 5 shows the earliest machine of this kind, and interesting comparisons may be made between that and machines in subsequent pages.

Among the simplest types of pillar machines are those which comprise in the main a lathe type of headstock, cast with or bolted to the pillar, and having a compound table on the knee to carry the work. There are three movements to the table—that in the vertical direction with the knee, one towards and away from the spindle, one transversely to the spindle. A few machines of this class have hand feeds only for all movements for the slides, through screws and handles, and through hand wheel and screw for the knee. Such machines have but a limited range, though with a dividing apparatus fitted to the table they are useful to brass-finishers and others for cutting hexagonal and other faces, and nicking screw heads, while boys can attend to them. Machines of this kind, too, generally have a screwed nose to receive chucks, the idea being that cutters can be turned up in place by means of tools held and operated on the table. There is no support to the outer end of the cutter arbor, and therefore neither long cutters can be used nor heavy tooling done. Those most generally employed are

Fig. 5.—First Universal Milling Machine built. (Exhibited by Brown & Sharpe, Paris, 1867.)

fly cutters, or of the face or end type, and as these lie close up to the headstock, they are stiff enough for the work which they have to do. Generally such machines are single-geared only.

For die sinking, special pillar and knee machines are fitted with vertical spindles, driven by belt through guide pulleys at the back. In such types the movements of the table are generally by hand, or the longitudinal traverse may be self-acting. A circular table is also sometimes fitted with a rotary motion through worm gear.

These machines are not standardised forms, but are generally of a somewhat special character. The simplest standard machines of pillar and knee type are those in which the spindles are horizontal, but in which the outer end of the arbor is steadied by an arm. These are generally termed "plain" machines, to distinguish them from the universals, which are built upon the plain types by the addition of an angular movement to the table, and of dividing heads and change wheels. On the plain machines rectangular relations only are obtainable. The methods of actuating the differents movements will vary in considerable degrees, rendering these movements of a more or less automatic character, but the range of the operations is still limited rigidly to rectangular planes.

A pillar milling machine, to be thoroughly complete and efficient must fulfil many conditions—conditions which are combined in one by the best manufacturers. Good useful machines at moderate prices are, however, made by many firms, and the low-priced ones will often be as well adapted for some classes of work as the more complete but higher-priced tools. In this as in other matters every tool user must consider his own special needs and requirements and pocket.

The leading dimensions of pillar machines are :—Distance from centre of spindle to the overhanging arm, which limits the diameter of the mills that can be used; the distance from the end of the spindle to the centre in the overhanging arm, which governs the length of mills that can be used. The height from the centre of the spindle to the top of the table, when the latter is in its lowest position, which limits the height of the work that can be tooled. The size of the table, and the length of longitudinal and transverse feeds, that regulate the dimensions of work that can be operated on. The terms "longitudinal" and "transverse" movements are used somewhat loosely by tool makers, the same terms sometimes indicating movements of opposite character. It will be better to state precisely the direction in any case. Properly, longitudinal traverse denotes the movement of the table at right angles with the spindle; and the transverse movement, that to or from the pillar.

The range of work which can be done on a plain milling machine, though usually restricted to rectangular movements only, is yet wide enough for nearly all the requirements of some firms; those, for example, who have no heavy milling, or gear cutting, or spiral-cutter making, or form milling. The work of the plain machine is done mostly by cutters threaded on an arbor. The diameter of arbor will range between about ½ inch and 1¾ inches in the smaller and larger machines respectively, and the length of arbor from about 10 inches to 24 inches respectively. The functions of the plain pillar machine lie chiefly in axial and edge work, done in strictly longitudinal planes. End mills can be used by supporting the arbor at the headstock end simply. It is convenient sometimes to be able to use an end mill, but such work is not as a rule so readily done thus, as it would be on a vertical spindle machine. Parallel grooves of tee or of plain section can be cut thus, and a series also of parallel grooves, by utilising the vertical movements of the knee. The principal utility of the machine consists, however, in the facilities which it affords for face and edge milling, using single cutters or gangs of cutters of plain and of profile outlines.

The principal points which are essential to the stability and good working of a pillar milling machine are, first, a substantial framing, well spread out at the base, to counteract the top-heaviness of the structure, and the one-sided arrangement of the knee and table; a large well-fitted spindle, with provision to take up wear, and bearings made of durable material; sufficient support to the outer end of the arbor, to maintain it steadily while cutting is being done—an important fitting, respecting the details of which opinions and practice vary. Another essential condition is the perfect alignment of the centre in the arm which enters the outer end of the arbor, with the centre of the spindle, in any position, whether closer in or farther out from the headstock. This means very perfect, close, and parallel fitting of spindle and of arm. The arm must be so fitted and held that it can be easily moved, yet firmly clamped in any position, or swung out of the way or removed altogether. It is essential, too, that the knee or table and the slides have large bearing surfaces, be fitted in the best manner, and in very accurate rectangular relations to each other and to the spindle, and with provision for taking-up wear. These points are all essential in the plainer, cheaper machines, as in those more elaborately fitted. They are details which are not apparent at a glance, and hence must be made matters for test, or,

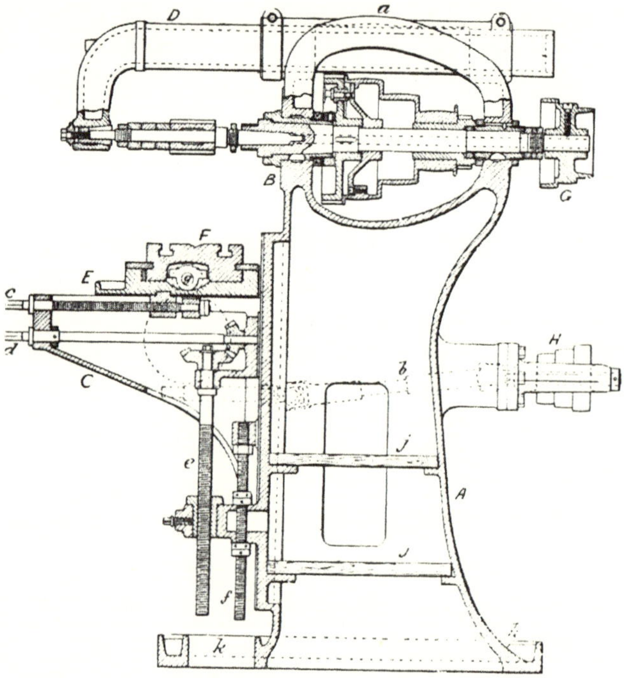

Fig. 6.—Vertical Section through Pillar and Knee Machine.

which is generally sufficient, taken on the guarantee of a high-class firm.

That which may be fitted, or not, to a plain machine, according to individual requirements, is provision for movements of a more or less automatic character. It is seldom that the longitudinal traverse motion is effected by hand only, since the addition of cones, a telescopic shaft, and gears is a simple matter. But in a number of machines the automatic movement is restricted to this alone. Another detail relates to the reversal of the table, which may be by hand only, or made automatic, with quick return. The use of adjustable stops in connection with this reversal, and, in fact, with all movements of the slides, is inseparable from perfectly equipped machines. Related to this closely is the need for micrometer readings. These are fitted to the bosses of the operating handles, so that exact depths of cutting can be done. A wide range of feeds also is necessary for the longitudinal traverse of the table.

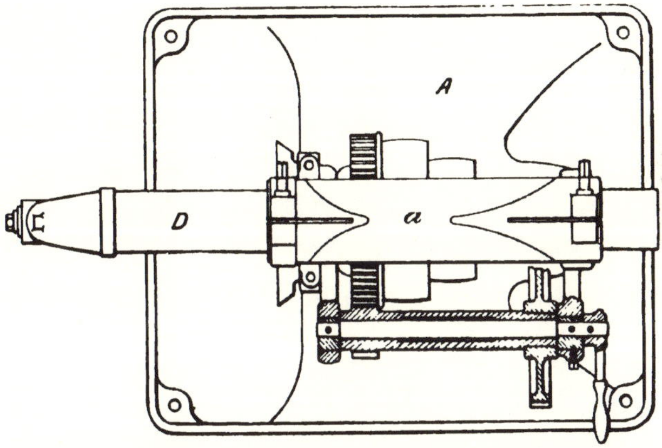

Fig. 7.—Plan, and Part Section of Pillar and Knee Machine.

Fig. 6 is a vertical section taken through a machine of the pillar and knee type, such as may, in its general features, be seen in large numbers of shops; and Fig. 7, a plan view, with knee and other details omitted. It is a type which is being subjected to much modification, particularly in the matter of feeds (see page 32), while the details of construction adopted by different makers vary in most parts of the design. The broad characteristics which distinguish all these are the horizontal spindle, the height of which is fixed, and the knee, which is capable of vertical adjustment on the pillar, or column. It is therefore a type of an exactly opposite kind to that of the Lincoln machine, with spindle vertically adjustable, and a table at a fixed height. The elements of the machine are as follows:—

The pillar A, of hollow or boxed section, and cast usually, as shown, in one with the headstock B, and having slide faces to receive the knee C. In a poorly designed machine, and in all in some degree, the head and the knee are the weak elements, the arbor being insufficiently supported, and the knee being liable to spring under heavy or wide cutting. Hence the reason for the fitting of the overhanging arm when edge milling is being done, and the fitting of bracings rigidly connecting arm and knee (see also page 86).

In the figure the upper part of the headstock is made continuous, forming a long boss *a*, Figs. 6 and 7, instead of having separate bearings, which is the case in many machines, particularly in the older designs. In the boss, the overhanging arm D is clamped by screws passing through split lugs. This arm is an important element, and is varied in design. It cannot be used for face milling, and is thrown up out of the way when this is being done. It receives a bushing to support one end of the arbor, in the example shown. Often a bracket is employed instead, sliding along a plain parallel bar, and the arbor is supported in that. Either arm or bracket is adjustable through the boss *a*, of the headstock, to receive arbors of different lengths. The advantage of the straight arm is that it need not be turned up, or removed when some adjuncts of the machine are put on the table.

The main spindle is generally back geared, as indicated in both Figs. The knee C carries the cross slide E, and the table F, the first having a transverse cross movement, or one to and from the pillar; the second a longitudinal, or traverse one, that is tangentially to the pillar. The mechanisms by which they are operated are varied greatly. Hand feeds are employed, and power. Power feeds are derived from the three-stepped cones G and H, driving in this example from the main spindle to the telescopic shaft *b*, which is jointed thus in order to permit of its following the vertical movements of the table. The objections to this method of feeding are stated on page 31, and substitutes for it are illustrated and described in some detail. *c* is the screw for cross traverse, or transverse movement, *d* that for vertical feed when effected by hand, which actuates the screw *e*, through bevel wheels, also used for power feeding. The screw *f* forms an adjustable stop for the height of movement of the knee. The table F is traversed by a screw, indicated at *g*, in universals, but often by a rack in plain machines. The advantages of a screw are finer precision in operating, and that the table remains at a standstill immediately the feed is disengaged. A rack permits

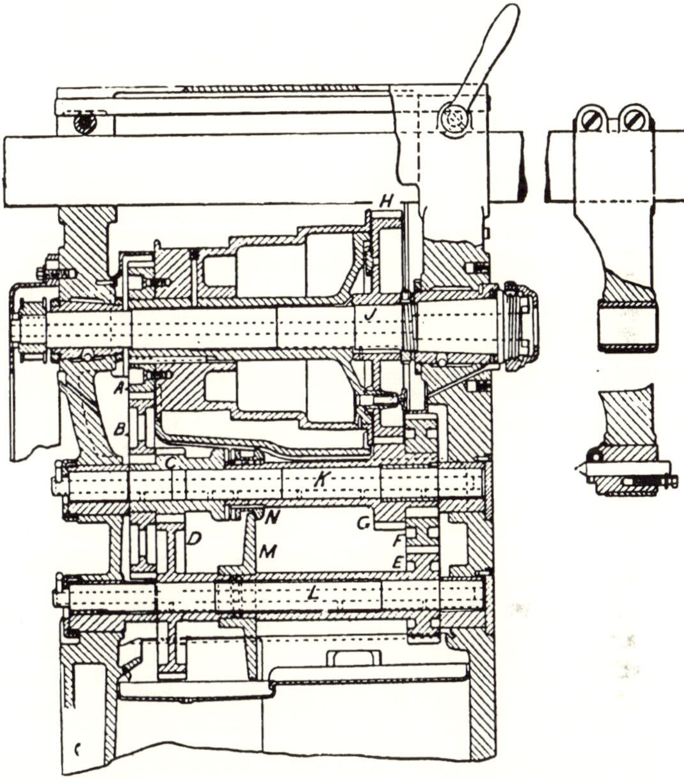

Fig. 8.—Vertical Section through Spindle and Back Gears, No. 5 Plain Milling Machine. (Brown & Sharpe Manufacturing Co.)

of making more rapid adjustments by hand when setting, and removing work. A universal also has a swivel table above F, which introduces other details. A universal head (see page 63) can be used on plain machines as well as on universals. But its utilities are limited on the first-named, spiral gears for example, requiring a swivelling table. Plain machines are also generally built stiffer than the others, and are sometimes termed manufacturing machines, to denote their greater power.

Other points to note in Fig. 6 are the waste-oil trough cast round the base of the pillar, and the shelves *j*, *j* within it for receiving tools, also the hole *k* cast in the base to permit the screw

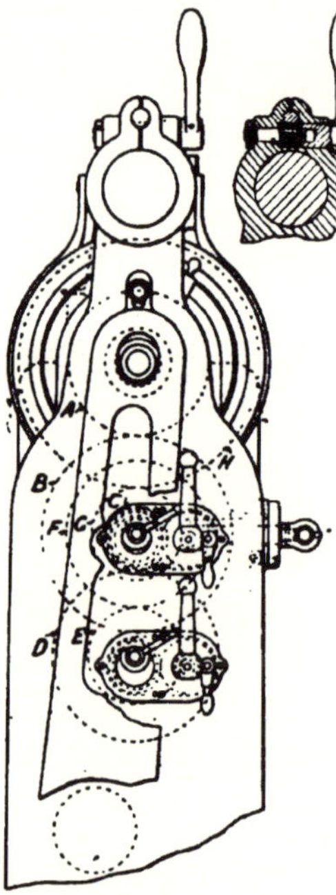

Fig. 9.—Rear View of Fig. 8.

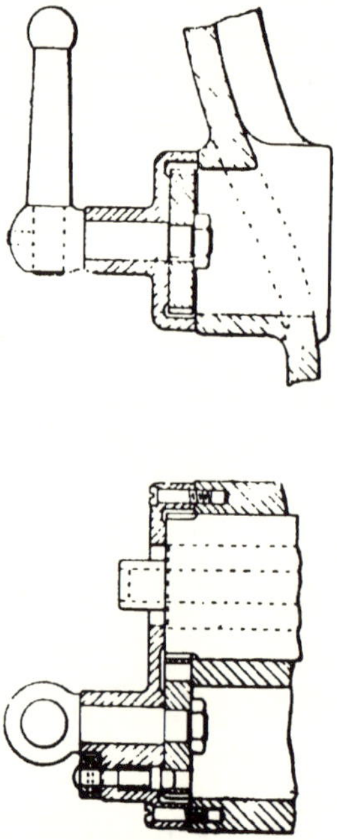

Fig. 10.—Sectional Detail of Eccentric Movements and Locking Pin in Figs. 8 and 9.

to pass down through the floor as the table is lowered This is not a good device, and an example is given of a telescopic screw (page 53) to obviate this objection.

Headstocks.—Figs. 8-10 show the headstock details of the Brown & Sharpe largest size of plain milling machine.

Fig. 8 shows the headstock in vertical section, with the back gears placed below, where they are covered in and protected. In the arrangement shown, the drive takes place from the cone pinion A to B, on an intermediate spindle, then through C, D, E, F, and G to H on the main spindle. The ratio thus obtained is 13·3 to 1. Both the back-gear shafts K and L have eccentric movements. On the gear quill on the shaft L there is a disc M, which, in the position shown, prevents the clutch N from being thrown into engagement. But when the gears D and E are thrown back and out, by the eccentric spindle L, that movement throws down the disc also, so permitting the clutch N to be engaged. When this is in, the drive takes place from the pinion A, through the gears B and G to H, with a ratio of 3·677 to 1. The upper gears can be thrown out also by the eccentric movement of K, giving a simple belt drive. By means of a two-speed countershaft, and the three-stepped pulley, and double-back gears, eighteen different speeds are available, ranging from 10 to 403 turns per minute, in geometrical progression as follows:—

SPINDLE SPEEDS IN REVOLUTIONS PER MINUTE.

Without Back Gears.

403	325	262	211	170	137
C. S. Speed Fast.			Slow.		

With First Back Gears.

110	88·5	71·3	57·5	46·3	37·3
C. S. Speed Fast.			Slow.		

With Second Back Gears.

30·4	24·4	19·7	15·9	12·8	10·3
C. S. Speed Fast.			Slow.		

The illustration, Fig. 11, shows the new headstock by the Cincinnati Milling Machine Co., designed mainly with a view to the employment of cutters of high-speed steel. The features present are, that a high belt speed can be used in conjunction with the power of back gears, and that the latter are double, providing a large range of spindle speeds, eighteen in number, as follows:—11, 13, 16, 20, 25, 30, 36, 44, 54, 66, 81, 99, 119, 146, 178, 218, 267, and 326. The countershaft speeds are also high, having 145 and 260 revolutions per minute. The smaller cone steps are larger

Fig. 11.—Headstock of Cincinnati Milling Machine.

than usual, for greater power. The steps measure $9\frac{3}{4}$, $10\frac{7}{8}$, and 12 inches diameter respectively.

The action of the back gears, seen in Fig. 11, is as follows:—The gears A and B slide in unison with a quill, which slides on the back-gear eccentric quill. A gears with C, producing slow driving from E to F. When B is slid into engagement with D, E drives F quickly. To bring either set into gear, the eccentric shaft must be thrown out, and the gears thrown in. The rim seen on B prevents it from engaging D when A and C are driving. When the machine is in operation, the back gears are encased.

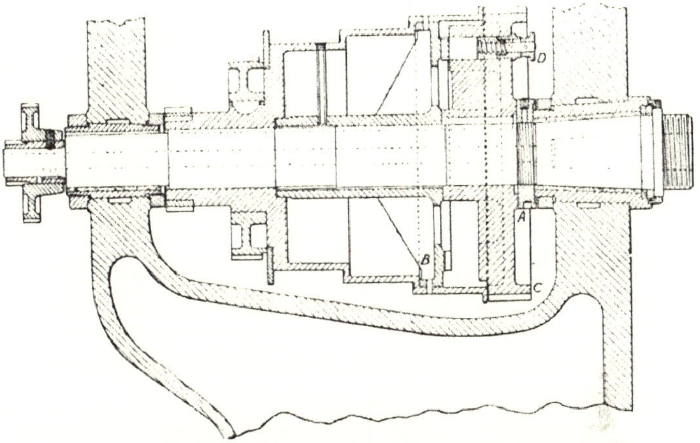

Fig. 12.—Headstock Spindle of Cincinnati Milling Machine.

Fig. 12 illustrates the headstock spindle of the latest Cincinnati machines, having the double back gear shown in Fig. 11. The spindle is hollow, of forged crucible steel, running in adjustable bearings. The hinder, parallel bearing has two lock nuts for adjusting diametrical wear. The front coned neck has but one. End play is taken up on this by turning the lock nut A on the spindle to the right, or in the direction in which the spindle turns, so drawing the latter into a close bearing. The clamping plate B for the spindle gear C is made separately from the cone pulley, and is fitted into the latter

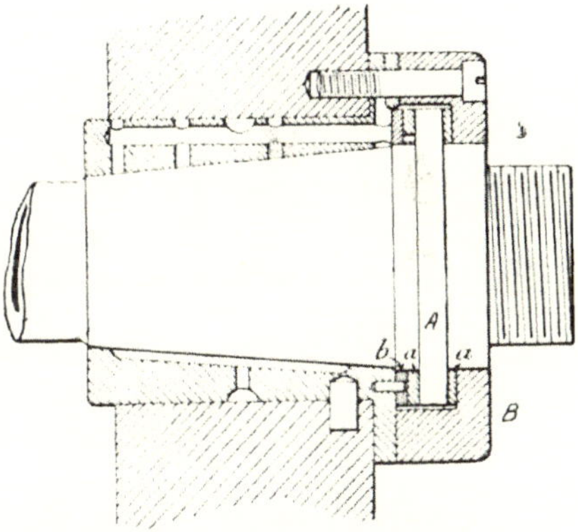

Fig. 13.—Garvin Front Spindle Bearing.

at front and back. The gear C is connected to the cones by the spring plunger D. Other details, as the keying of the back gears

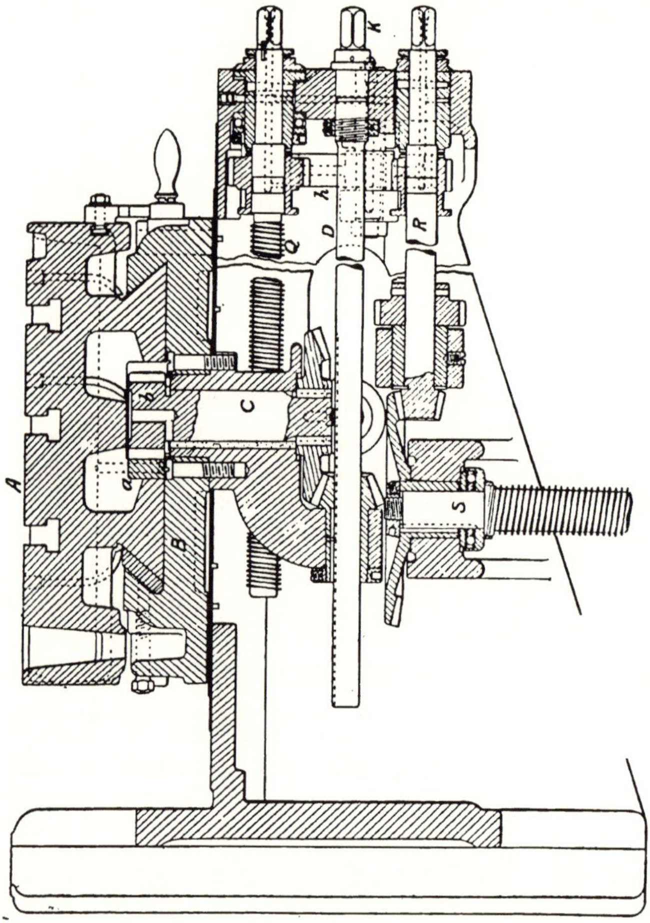

Fig. 14.—Section through Knee and Table, No. 5 B. & S. Miller.

on the cone boss to the rear, provision for lubrication, the protecting cap to the front bearing, &c., will be noticed.

Fig. 13 illustrates the Garvin form of front spindle bearing. The neck is tapered to an angle of 5 degrees, and end play is prevented by the washers which come against the collar A.

Two of these washers *a*, *a* are hardened and ground, the third *b* is a soft one, which can be taken out and its thickness reduced in the lathe by the amount to which end play might have developed. The washers and spindle are held in contact by the cap B, screwed in front. Lubrication is provided for by the groove and channels seen. The hinder bearing is parallel, with a bushing tapered externally.

Other details of spindles and bearings will be found on pages

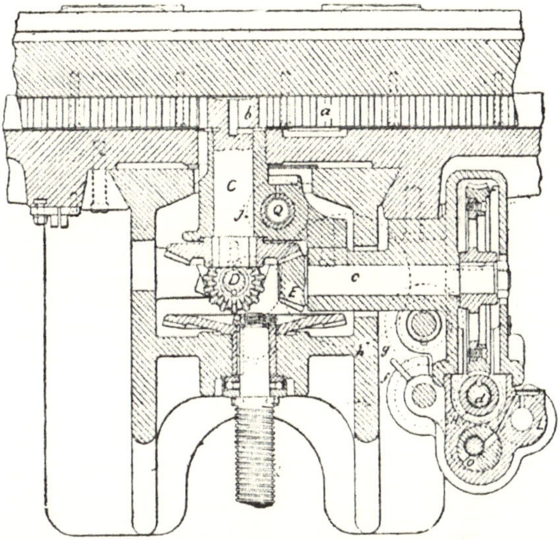

Fig. 15.—Transverse Section through Knee and Table, No. 5 Miller.

34, 36, and 39, in connection with illustrations of feeds. In the meantime we consider the details of knees and tables.

Table and Knee Feeds.—Figs. 14-19 show the table and knee details of the Brown & Sharpe plain miller. In Fig. 14 the fitting of the table A to the cross slide B, on the knee, is shown in section, with the take-up strip. The table is rack driven, as is usual with plain millers, the vertical rack being indicated at *a*, compare with Fig. 15, and the pinion at *b*, on the short vertical shaft C. This is actuated by hand from the front of the machine by the shaft D, through the bevels seen, and by power from the pinion E, driven from worm gearing, the worm wheel F being on the opposite end of the shaft *c* that carries E.

The feed is automatically tripped by the lever G, Fig. 16, without dropping the worm H, which is a frequent device, and the objection to which is that it leaves the table unlocked. If the worm is allowed to remain in gear the table remains fixed in its place whenever stopped. But it is necessary to disengage the worm when the table movement has to be effected rapidly by hand, and this is done by moving an eccentric J, Fig. 16, that throws down the hinged bearing in which the spindle *d*, of the worm H, is carried. A crank handle in front of the knee, at K, operates the quick feed through the bevel wheels on shafts D and C. There is also a fine hand feed to the table, with graduations reading to $\frac{1}{1000}$th of an inch, operating the worm gear through an intermediate pinion L, Fig. 15.

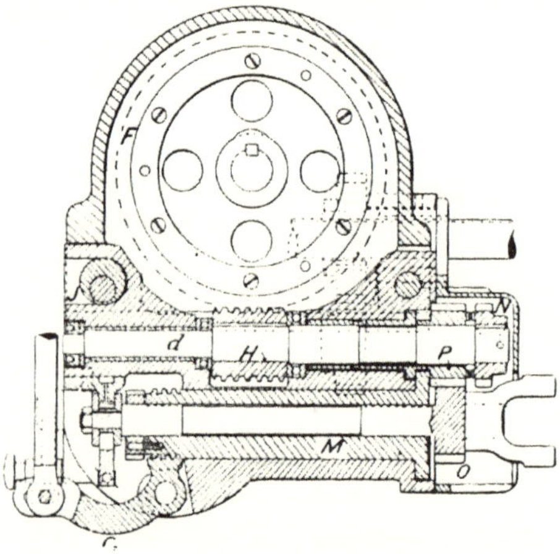

Fig. 16.—Vertical Section through Worm Gear, &c., driven by Telescopic Feed Shaft, No. 5 Miller.

When the automatic feed is tripped, it is done through the lever G moving an eccentric sleeve M bodily along, and so disengaging the clutch N, Fig. 16. This clutch is constantly running, being in the train of connections between the sprocket chain drive from the spindle to the box of feed gears, and thence through the universal joint to the table feeds just described.

The eccentric sleeve M, which is connected with or disconnected from the clutch, is employed to reverse the table feed, engaging either directly from pinion O to pinion P, or through the intermediate L, for reversal, being operated by a lever.

The cross and vertical feeds can be operated by power, or be disconnected. The worm H furnishes the drive in both cases.

For the cross-feed screw Q, the drive takes place from pinion P, through *f*, *g*, *h*, and *j*,—*j* being driven from the splined clutch on Q.

The tripping arrangements of the table are shown in the detailed drawings, Figs. 18 and 19. The method of elevation of the knee by hand, or power, is clearly shown in Fig. 14, R being the horizontal spindle driving the vertical feed screw S through bevels, the mass of the knee, &c. being carried on a ball race.

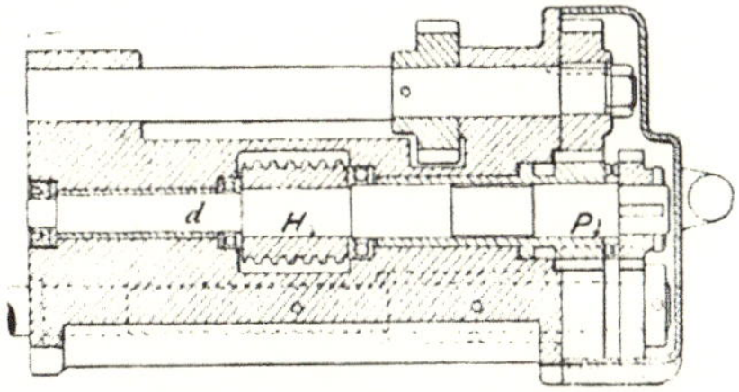

Fig. 17.—Horizontal Section through Worm, &c., driven by Telescopic Feed Shaft, B. & S. Miller.

A good deal of discussion has taken place on the question of the best method of imparting milling machine feeds. The old plan was, and it is still the one chiefly adopted, to drive the feeds directly from cones on the end of the main spindle (see Fig. 6, page 20). The result is that only a medium range of workable feeds is available when using the medium - spindle speeds, while at the extremes of small cutters it is impossible to get feeds at all suited to the speeds at which those cutters should be run economically. Using a large cutter, requiring a slow-spindle speed, the feed will be much too slow. Using a small cutter requiring a high-spindle speed, the feed will be much too fast. The case has been put by Mr P. V. Vernon thus:—

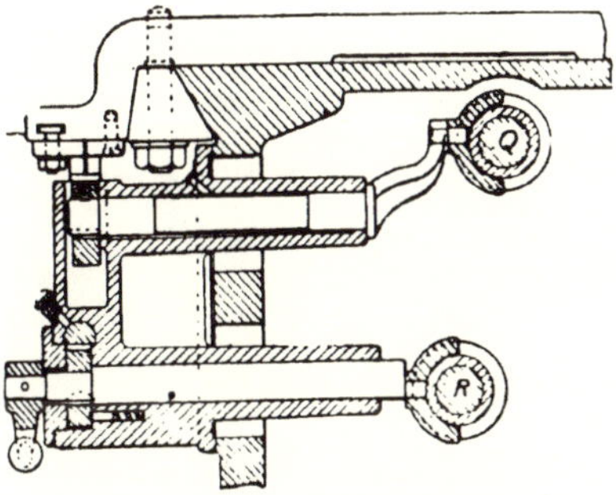

Fig. 18.—Trip Movement, B. & S. No. 5 Machine.

"Assume spindle speeds varying from, say, 10 up to 300 turns per minute, giving a 30 to 1 ratio of highest to lowest speeds.

"Assume, as a suitable range of feeds for slowest spindle speed, $\frac{1}{2}$ inch up to 8 inches per minute, giving a 16 to 1 ratio of highest to lowest feeds. (The above ranges both of speeds and feeds are well within the usual limits.)

"At the highest spindle speed (300) the above feeds would become 15 up to 240 inches per minute, as they would increase with the speed of the spindle, and would of course be quite useless on the higher speeds.

"Conversely, if the range $\frac{1}{2}$ inch up to 8 inches per minute could be obtained on the highest spindle speeds, these feeds would be only $\frac{1}{30}$ of that amount on the slowest spindle speeds, and would also be of very little use, the highest feed for the slowest spindle speed being only about $\frac{1}{4}$ inch per minute.

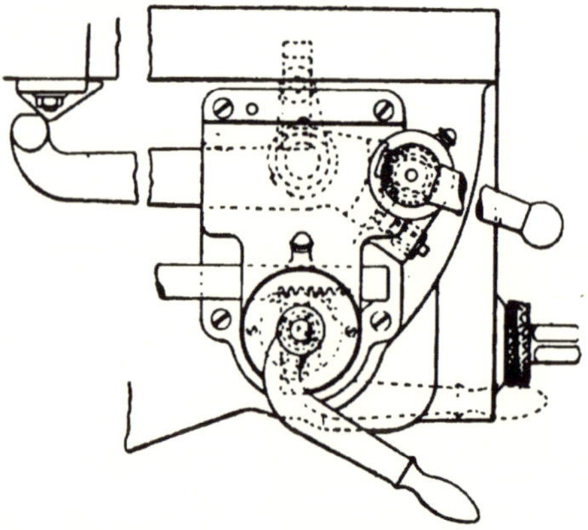

Fig. 19.—Trip Movement, B. & S. No. 5 Machine.

"In order to obtain the proper range ($\frac{1}{2}$ up to 8 inches) at all spindle speeds it would be necessary to have a total range of feed variation provided by means of the change mechanism giving a total ratio of (30×16) to $1 = 480$ to 1, which is obviously impracticable."

The case is thus very strongly put for the driving of feeds from the countershaft, or if from the spindle, in such a way that they shall not respond to the variations in spindle speeds, but be driven from a speed independently capable of variation to suit the sizes of the whole range of cutters being used.

Numerous designs have been fitted which provide independent feeds by means of gears—hence termed positive feeds. An objection to these is that they will not slip in case of slip of the main belt occurring, and the feed going on, is likely to cause fracture of something. A slipping clutch, or breaking piece, is therefore sometimes fitted in the feed-gear trains.

But the question of belt versus geared feeds is still in a state of transition. Many prefer the belt because it would slip rather than cause a jam. There need not be much virtue in gears over belts in point of power, though the belief is usually tacitly accepted. That is strictly true only when comparison is made with the narrow belted feeds hitherto used. A belt feed can be made powerful, and a sufficient number of variable feeds obtained therefrom, provided it is not driven from the spindle speeds, but from the countershaft. Though it is not possible to get a powerful feed from a slowly rotating main spindle, it can be obtained from the countershaft.

Whichever method is adopted, the separate feed gear not only provides an independent range of feeds, but it simplifies calculations in two ways. It is easier to understand feeds irrespectively of varying spindle speeds, than in connection with them. Also it takes account of the diameter of the cutter, or number of teeth, instead of feeds per revolution, which does not consider cutter diameter, or number of teeth.

It may also be pointed out that economy in power is studied by separating the feed drive from the spindle drive. When the latter drives the former, power is taken directly from the spindle, which would be better utilised for heavier cutting, unless there is excess of belt power.

One advantage of the geared feeds in the heavier machines is that they are better able to feed heavy knees, carrying heavy work vertically, than the ordinary belt feeds are.

The objection to an independent feed for milling machines, that is, one obtained by arrangements entirely disconnected from the main machine spindle, is that the spindle belt may slip under heavy duty while the feed continues in operation, with disastrous results. But the independent feed is nevertheless desirable, and it has been embodied in many machines of late years. Feed is given in inches per minute, or parts of inches per revolution, and this cannot be estimated directly, but only by reference to the spindle speed.

The desirability of having a table feed independent of the spindle speed induced Messrs Brown & Sharpe to design a machine with constant and uniform belt speed, and variable spindle speed, and variable table feed, derived from the constant-motion spindle. It has been in operation for some years, and given satisfaction.

The single-driving speed renders the machine adaptable to an ordinary belt drive, or motor drive with sprocket chain. The spindle speeds range from 15 to 376 revolutions per minute, arranged in geometrical progression. They provide, with a surface speed of cutter of 20 feet per minute, for a range of cutters from $\frac{3}{16}$ inch to 5 inches in diameter; with a surface speed of 30 feet per minute, for cutters from $\frac{5}{16}$ inch to $7\frac{1}{2}$ inches; and with a surface speed of 40 feet, for cutters from $\frac{3}{8}$ inch to 10 inches. The speed mechanism is as follows:—

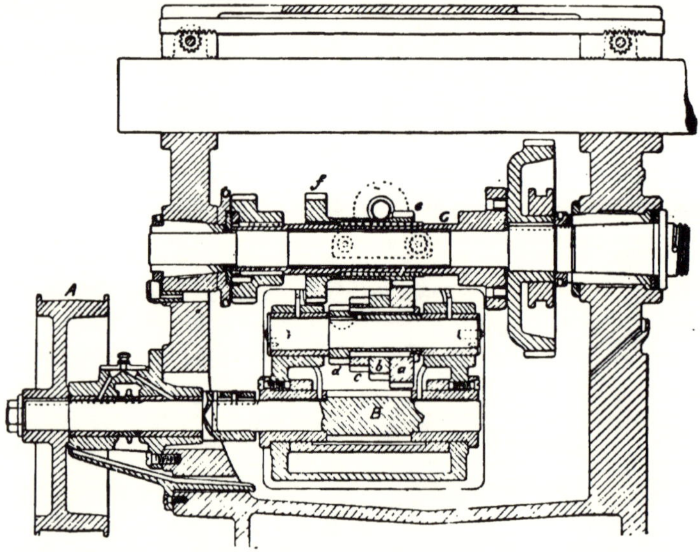

Fig. 20.—Vertical Section through Main Spindle, No. 2 Universal Machine. (Brown & Sharpe Manufacturing Co.)

The main driving pulley is seen at A, Fig. 20. This drives the long pinion B, which, through an idler C, Fig. 21, may actuate either *a*, *b*, *c*, or *d*. The lever D slides the idler C along to engage with either wheel in the series, and the lever E then lifts it into gear. The drive as shown is taking place from *a* to *e*, but a lever provides means for sliding *e* out of gear, and bringing the larger wheel *f* into engagement with the smallest wheel *d*. A fast or slow motion is thus obtained for any one of the combinations obtained between B and the cone of gears *a*, *b*, *c*, *d*.

Besides this, there is a back-gear set, derived from *c* and *f*, which wheels slide upon a quill G. This is shown in the next Fig. 22, where H is the back-gear spindle. This has a cam *g*, which actuates a forked lever, and this in turn a collar J, connected to locking pins *i*. The handle K engages the back gear, the first action of which is to withdraw the pins *i*, and the final action is to throw the collar J to the left. This causes compression of the springs behind the pins, which, on the first turn of the machine, throws the pins into place with a sharp snap.

The guard which encloses the back gear is cut away as shown, Fig. 22, to permit the quill, encircling the eccentric spindle H, to be turned by the hand, in order to bring the gears into a position to engage, and the outside of the quill is knurled with that object.

A small gear is shown at *k*, which can be made to engage with the main back gear M by pulling out the knob *l*, but at other times it is kept clear of the wheel by the encircling spring shown. The reason for this is to permit the spindle to be turned by hand when the drive is by motor, as in that case the sprocket chain with its wheels is enclosed by a shield.

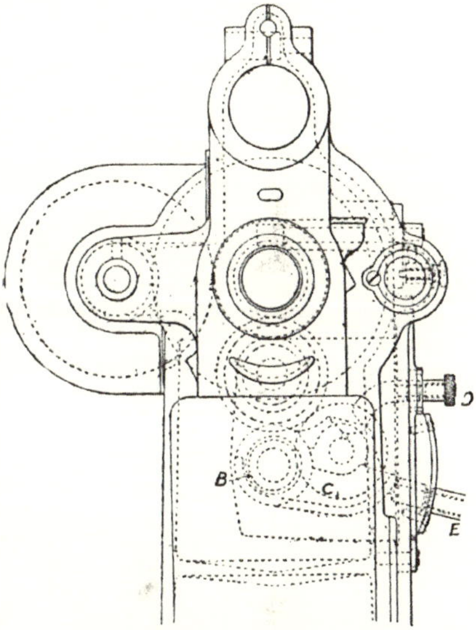

Fig. 21.—End View of Headstock, B. & S. Machine.

The table feed is rendered independent of the spindle drive as follows (Fig. 23):—

The sprocket A, driven by a pitch chain, drives the long pinion B. An idle wheel, not indicated, but which is similar to C in Fig. 21, and carried by the frame D, engages with either *a*, *b*, *c*, *d*, *e*, or *f* by the lever E, after its position has been located by a lever similar to D, in Fig. 21. In the figure, *b* engages with *g* on the telescopic feed shaft F; but by reversing the lever G, *g* is slid out, and *h* put into engagement with *f*, so giving choice of fast or slow feeds for either one of the combinations of the cone of gears, as in the speed drive just noted in Fig. 20.

Gear plates are provided for speeds and feeds, and are placed directly above the speed and feed-changing levers. The speeds given for each position of the stop pin of a lever are marked immediately above the stop-pin hole for that position.

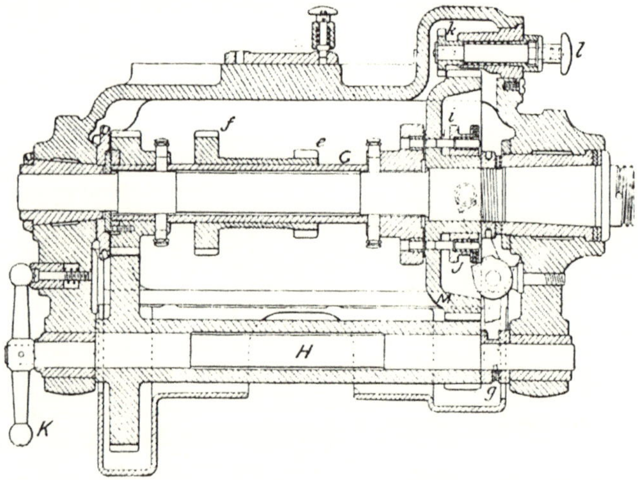

Fig. 22.—Horizontal Section through Back Gears, B. & S. Machine.

Some other feed arrangements applied to the Brown & Sharpe machines are shown in Fig. 24. The feeds are derived from the spindle, driving down to the sprocket wheel A, in the first place. The shaft B is thus rotated. B carries two wheels, C and D. C runs on B, and has a long boss which forms a journal for wheel D. These wheels are engaged by the clutches *a* and *b*, operated by a lever outside the gear box, the range of movement being controlled by a knob. C operates a slow series of feeds, D a series of fast ones. These first movements are transmitted through the wheels on the intermediate shaft E to the wheels on the feed shaft F, whence the table is operated through a telescopic shaft. The details are as follows:—

The two nests of gears G, H are keyed upon their shaft E. The gears J, K are loose on their shaft F. The latter wheels have long sleeve bosses, and either one in the nest is engaged with its fellow on the shaft E by means of a series of six locking pins, two of which are shown at *c* and *d*, and which enter recesses in the

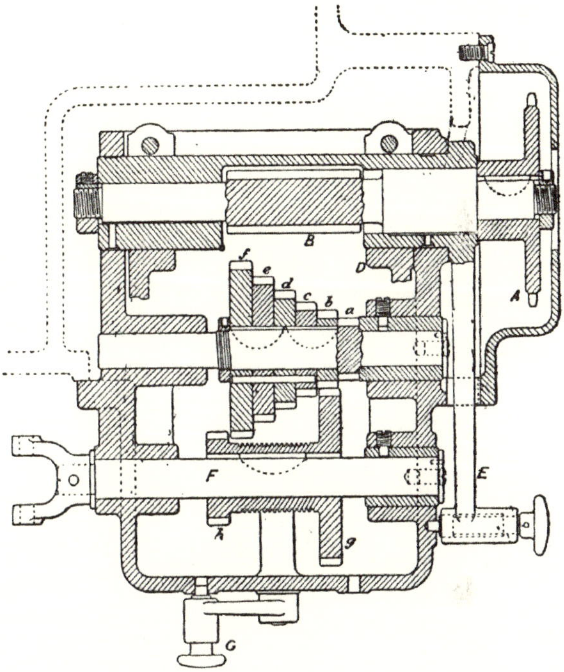

Fig. 23.—Section through Feed Gears, B. & S. Machine.

bosses of the gears. Their positions are controlled by an index disc *e*, that turns on the disc *f*, and carries a cam for actuating levers which are fastened to the ends of small pinions that engage with rack teeth (not shown), cut in the locking pins. The pinion levers drop into a recess in the index disc, and as there is but one recess, only one feed can be engaged at a time.

The feeds of the recently designed motor-driven pillar and

knee machines of the Cincinnati Milling Machine Co. are shown in section in Fig. 25. The machine is driven by a variable-speed motor, having field control. The entire range of speeds is obtained by the motor, supplemented by two sets of back gears. The total range gives a surface speed of 20 feet per minute to cutters of from $\frac{5}{16}$ inch to 6 inches diameter, or of 40 feet per minute to cutters of from $\frac{5}{8}$ inch to 12 inches diameter. Power is taken from the motor to a friction clutch in the first place, and thence to a sleeve on the main spindle. The friction clutch is introduced into the

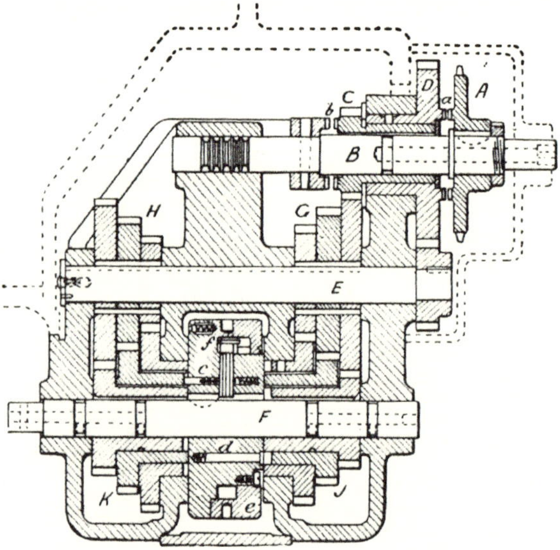

Fig. 24.—Another B. & S. Geared Feed.

drive in order to permit of instant starting and stopping of the machine without waiting for the motor either to stop or to attain full speed, which must be done when the drive is effected directly from the motor. The spindle can also be turned by hand on disconnection of the clutch. The field rheostat is not disturbed when starting or stopping the machine, whether done by the friction clutch or the main switch. It is set once for all to the required speed, and is not altered unless a change of speed is wanted.

In Fig. 25 the motor drives through a sprocket chain the

wheel A, which drives the sprocket B through the clutch C, operated by the lever D, coming within reach of the attendant. The second chain comes from B to the wheel E, which is mounted on a sleeve F. This wheel and sleeve either drive the spindle G directly through the usual pin *a*, or through the back gears H and J, or K and L. These are thrown

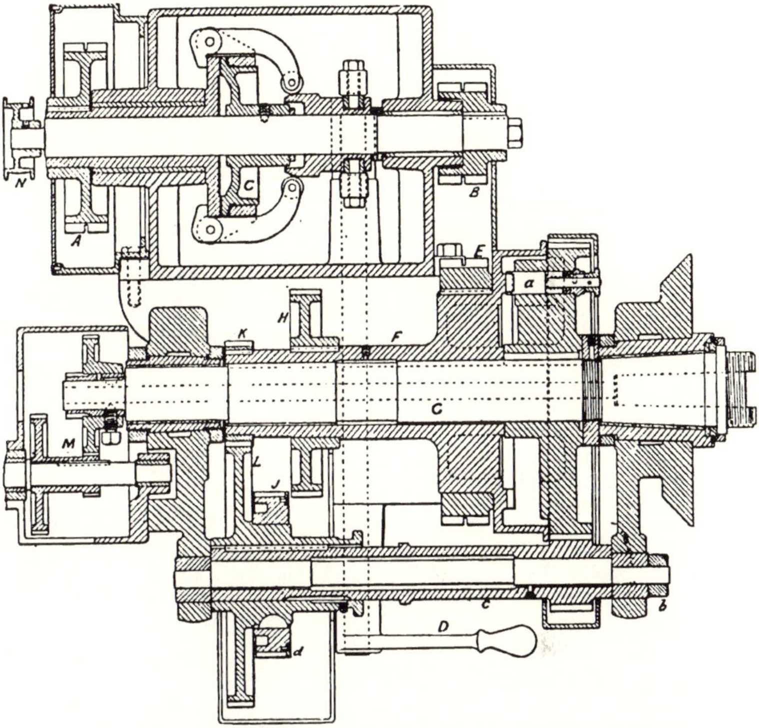

Fig. 25.—Section through Feed and Spindle Gears of Cincinnati Machine.

into or out of engagement by the usual eccentric spindle and lever, the boss of which is seen at *b*. The endlong movement is effected by sliding the sleeve of J and L along the main sleeve *c*. A flange or shroud *d* on the wheel J prevents risk of endlong movement taking place when either pairs of gears are engaged. The gears M transmit the feed motion. The small pulley N is for driving the oil pump.

The transmission of the feeds of the recent Cincinnati machines is obtained through the gears M directly from the rear end of the spindle. One box of gears is placed there, the other at the rear of the column, and the two are connected by a vertically inclined shaft. The first box is shown in section in Fig. 26. It provides by the sliding sleeve A, with its two pinions engaging with the pinions *a* and *b*, on the spindle end, two speeds for each cone speed. The sleeve is operated by a lever *c*, outside the encased box. Thence the mitre wheels B transmit motion to the feed gears in the box below on the column. In the latter are nests of

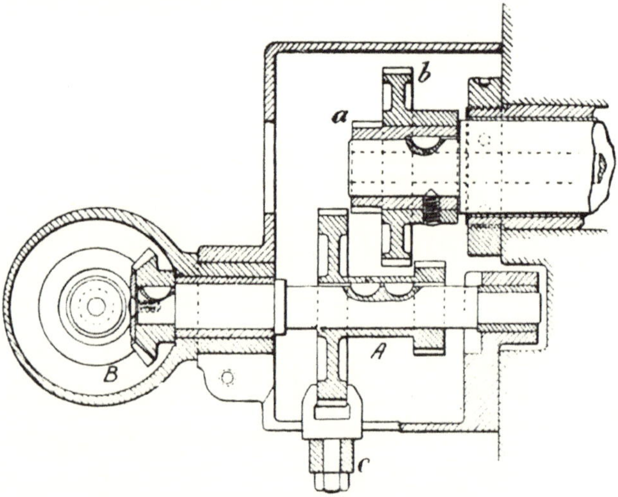

Fig. 26.—Sliding Gears at Rear of Spindle of Cincinnati Machine.

gears driven in the first place by two wheels, the common spindle of which is actuated by mitre gears from the telescopic shaft. Each of these two wheels is in permanent engagement with a wheel on one of two nests of spur gears below, arranged in cones in the manner now so common in feed gears. These cones of gears are mounted loosely on their shaft, and are independent of each other. The larger of the two upper wheels is in engagement with the smallest wheel on one set of cones, and the smaller wheel with the largest on the other set of cones, so that the two sets of cones are run at widely different rates. Variable rates are transmitted thence to the machine slides by an intermediate sliding gear on a separate spindle, which can be slid along and engaged with any wheel on either series of cone gears. There are two levers used in this box, one of which is set opposite to figures on a quadrant, that indicate the rate of feed in thousandths of an inch per revolution of the spindle. The upper lever next brings the intermediate wheel into engagement.

Fig. 27 illustrates a feed gear fitted to the Garvin machines for

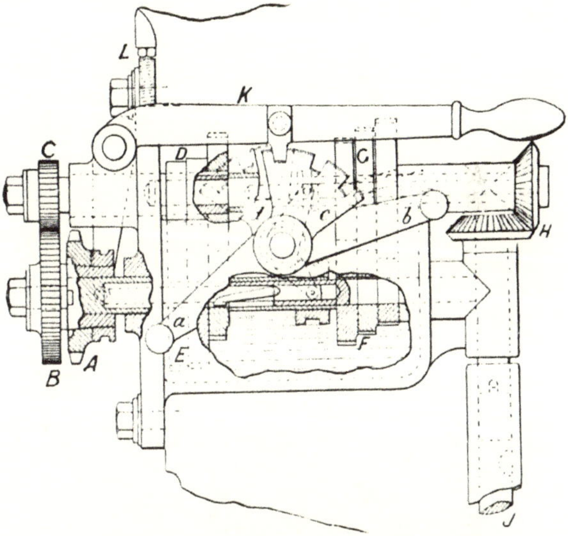

Fig. 27.—Feed Gears of Garvin Machines.

feeding the knee and its slides, with provision for a series of eighteen changes by the movement of handles.

The gears are enclosed in a box within the column or pillar of the machine. A portion of the gears projects at the back, being covered with a removable cap, and the operating levers are at the side adjacent within reach of the left hand of the attendant.

The power for feeding is transmitted from the main spindle to the sprocket wheel A, on the outside of the box. This engages by a sliding clutch with the spur wheel B to the left, which is in engagement with the wheel C above, on a short shaft, on which three gears, D, are keyed fast. These drive to either of the three wheels below, which are loose on their shaft, and are put into action by a sliding spring key seen in the broken section. The lower shaft carries also to the right, three gears, F, keyed fast upon it, which communicate the motion derived from either one of E, to three others, G, above, loose on a short shaft, which has no connection with that on which the wheels D are keyed. The motion of G is then transmitted to the knee by the mitre wheels H, and the feed shaft J.

For any one position of the left-hand key in the wheels E, there are three changes available by the right-hand key engaging with the wheels G, and since any combination of three can be made, nine total changes are available. And besides, as the gears B and C are reversible on their spindles—on the removal of the protecting caps (not shown)—the series of nine changes can be doubled. These range from $\frac{1}{270}$ inch to $\frac{1}{4}$ inch per revolution of the spindle.

Each sliding key is provided with a lever *a*, *b*, having a notched sector or dial plate *c*, *d*, for the right and left hand key respectively. The notches are numbered, and a table at the side of the pillar shows the combinations which give certain feeds. But before any change can be made, the locking lever K has to be raised, which movement disengages the clutched sprocket wheel A from wheel B, and so stops the movements of the gears, while the keys are being shifted, though the machine continues running.

The wheels are of hardened steel, and run in a bath of oil. The sprocket chain is tightened by the set screw L, which adjusts the box bodily in the vertical direction.

Fig. 28 illustrates a knee feed for the Kempsmith pillar and knee miller. It gives sixteen changes, ranging from 0·004 to 0·150 inch per revolution of the spindle. The gears are enclosed in a box at the side of the pillar, and are driven from the spindle through a sprocket chain to the wheel A. The sectional view is drawn as though the shafts were arranged in one plane, though they are not so actually.

The sprocket wheel A is keyed to its shaft B. A double-ended clutch C is capable of a sliding movement on this shaft, by the lever D, which is rocked by the pin of the eccentric shaft, indicated in end view. The clutch can thus be engaged with either the wheel E or F, which, it will be noted, impart either a faster or a slower motion to the cone of gears on the shaft G. The corresponding gears are mounted on the shaft H, in a swinging cage J,

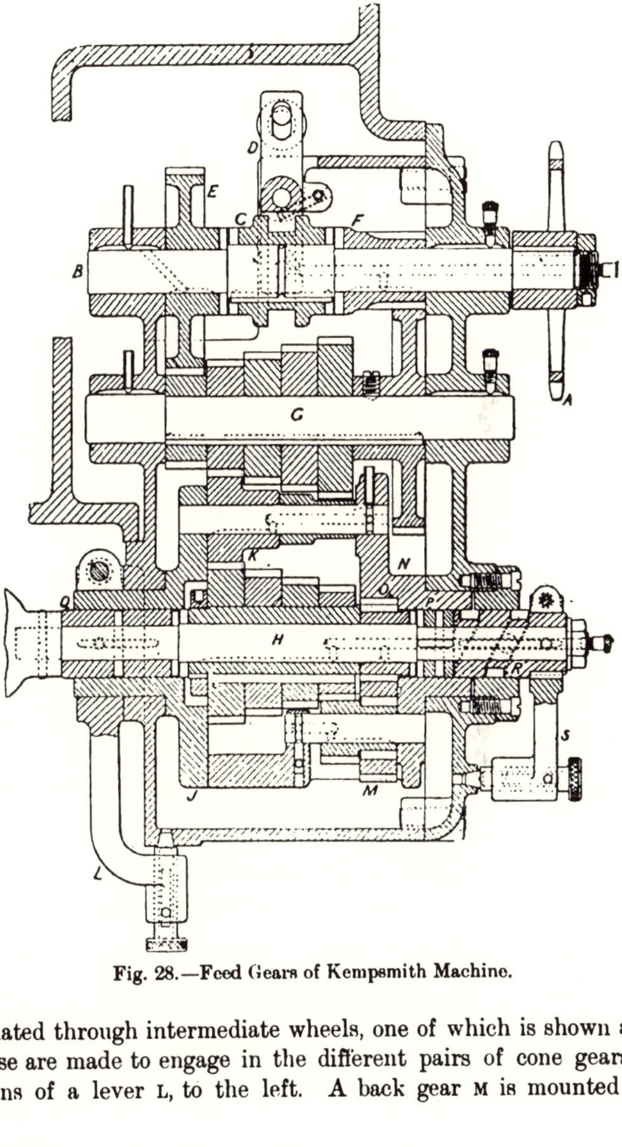

Fig. 28.—Feed Gears of Kempsmith Machine.

actuated through intermediate wheels, one of which is shown at K. These are made to engage in the different pairs of cone gears by means of a lever L, to the left. A back gear M is mounted in a frame N, and drives the pinion O; P and Q are clutches, keyed, and pinned to the shaft H. A sleeve R has a helical groove which is operated by a lever S, to the right, to engage either P or Q. The

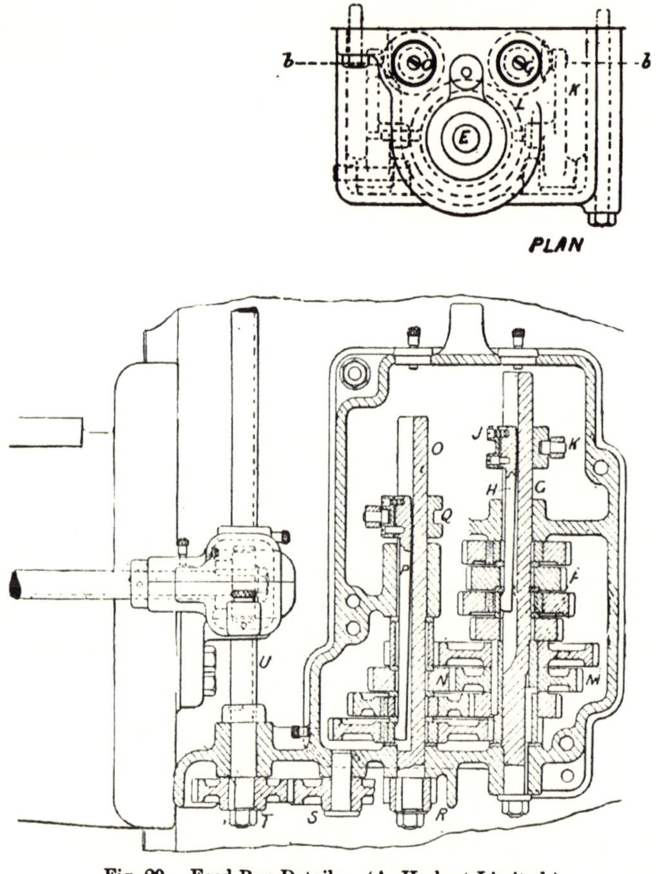

Fig. 29.—Feed Box Details. (A. Herbert Limited.)
Section in Plane *b-b*.

end of the telescopic shaft is seen to the left. A plate attached to the gear box indicates the position of the levers for any feed.

Figs. 29-32 illustrate a geared dial feed fitted by Messrs Alfred Herbert Ltd. to recent machines, Fig. 33. The views comprise plans and sections of the gear box attached to the main framing. The feature to be noted is that the simple turning of an index wheel A gives at once the rate of feed to which the pointer is set.

The nests of gears, entirely enclosed, are carried on three vertical spindles which revolve and carry the various gears. Tracing the feed movements from the shaft B, in the first place, belt driven, compare with Fig. 33; this drives mitres C, the second of the pair having a spur D on its spindle, which engages with a wheel on the shaft E. There are four such wheels on a sleeve running loose on shaft E, and each is engaged with the gears F on shaft G Either one of the latter gears is made to drive the shaft G, by means of the sliding spring key H, retained and pivoted in the

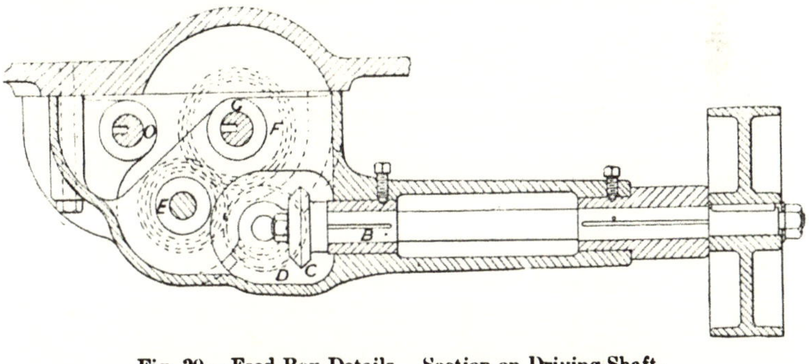

Fig. 30.—Feed Box Details. Section on Driving Shaft.

collar J, and actuated by the lever K. The method of moving this lever is the most interesting feature of the device, being effected by the movement of the index wheel A, which revolves a cam block L, having grooves cut around its body, which grooves coerce pins in the lever K, as seen in Figs. 29 and 32, and so rock the levers upwards or downwards according to the direction of rotation of the index wheel. The shaft G is thus rotated at either one of four different speeds, and imparts motion to four gears M, keyed on it through the medium of a sleeve forming part of the smallest gear. Then these four gears drive others N, on a shaft O, fitted up with a sliding key P, and collar Q, actuated also by a cam groove in the block L. Sixteen different speeds are thus obtained through the movement of the hand wheel A. Finally, the gears R, S, T convey the movement to the splined vertical feed shaft U, Fig. 29, which drives the horizontal feed shaft to the table through spiral gears.

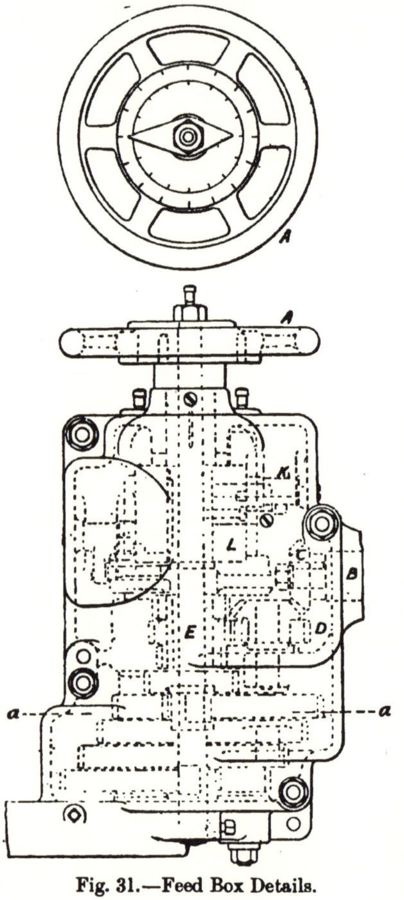

Fig. 31.—Feed Box Details. Front Elevation.

The machine to which the feed box just described is fitted is shown by the photo in Fig. 33. It combines several features of well-recognised value with some novel ones.

The principal novelty lies in the dial feed box just described, by means of which any required feed can be obtained by the simple movement of the hand wheel seen over the box in Fig. 33, the feeds being marked consecutively round the edge of a disc on this hand wheel, the dial being rotated until the number corresponding to the feed required comes opposite the pointer. No other movement whatever is required.

Another feature is that the feed can be driven either from the countershaft, or from the spindle as desired, the combination giving the most complete feeding arrangements. In another machine by this firm, the vertical spindle one, illustrated on page 108, the feed is driven from the countershaft only. Another detail which is also illustrated in that machine, as well as in Fig. 33, is that the usual telescopic feed shaft is abandoned in favour of gear drives through shafts at right angles.

Other valuable points in this machine are the following:—The knee is of boxed form, being completely closed underneath. Its bearing on the column extends above the plane surface of the knee, so conducing to steadiness of movement. Telescopic cover plates are fitted to prevent dust and cuttings from falling on the mechanism within the knee. The table is thick and massive, and is oiled from the side, so that fixtures and vices need not be disturbed for the purpose of lubrication. This is a point of much importance when a machine is kept a long time in service with the same fixture upon it, doing repetitive work. The screw by which the table is operated does not rotate, is always in tension, and is not splined. A covered channel in the table extends from end to end to enable the water used in lubricating the cutters to pass away freely.

The good practice of fitting hand wheels to move the slides in preference to using levers is embodied. The rim of a hand wheel is readily felt, while a crank handle has to be searched for round the

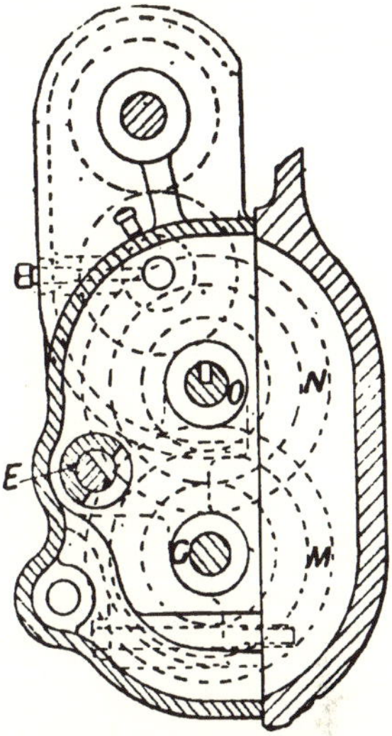

SECTION a-a

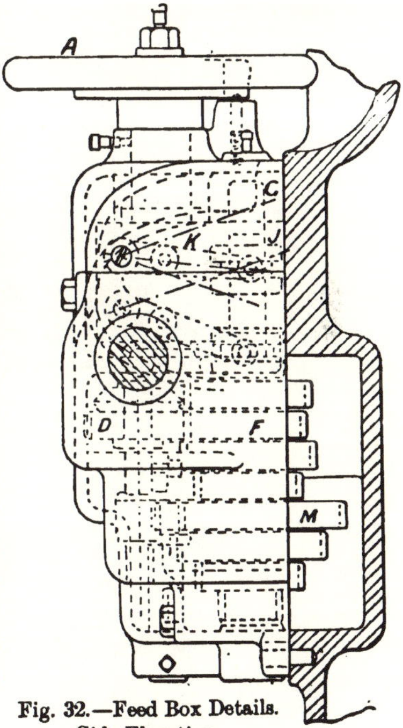

Fig. 32.—Feed Box Details. Side Elevation.

circle. Also when the knee is lowered, the attendant has not to stoop to a hand wheel, while he would often have to do so for a

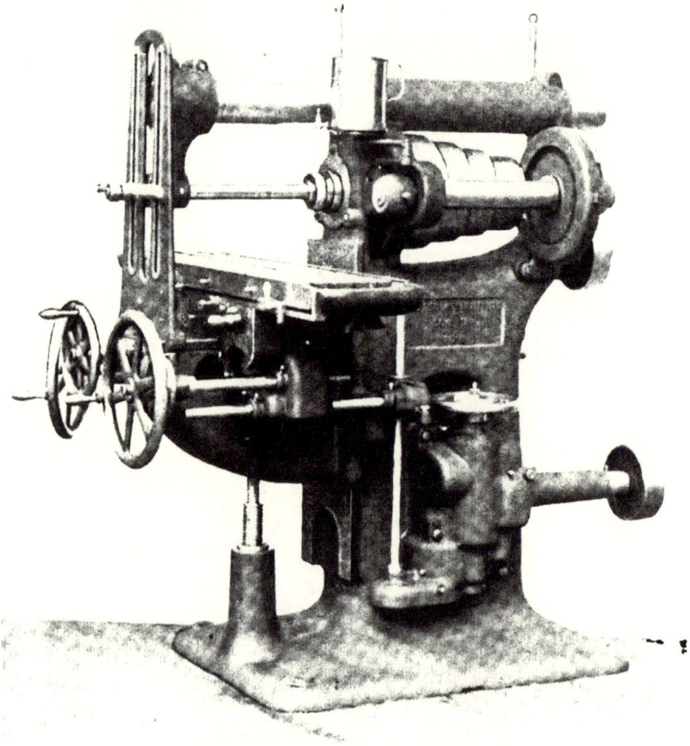

Fig. 33.—Milling Machine with Dial Feed.

handle. Another feature is that the feed shafts are separated sufficiently far to permit of having an independent hand wheel for each motion, so avoiding the trouble of changing wheels when making adjustments. These are apparently minor matters, but they tell up in the economies of the work of the machine. The elevating screw is of the telescopic type, so doing away with the necessity of cutting a hole in the floor. All motions of the slides have clamping handles. The feed has automatic trips, and dead stops. All motions have adjustable graduated index discs. The head fittings embody the best practice. The head, with its sleeve, is cast solid with the pillar. The sleeve is tubular, embracing the overhanging arm, and pinching it with clamping handles. The arm is of solid steel bar. The spindle, of crucible steel, is bored right through. Its nose is threaded, and protected with a cap

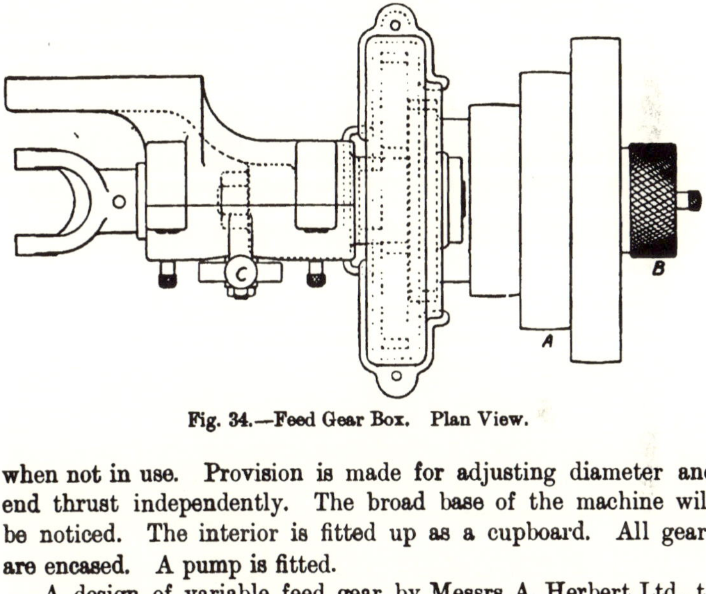

Fig. 34.—Feed Gear Box. Plan View.

when not in use. Provision is made for adjusting diameter and end thrust independently. The broad base of the machine will be noticed. The interior is fitted up as a cupboard. All gears are encased. A pump is fitted.

A design of variable feed gear by Messrs A. Herbert Ltd., to be driven either from spindle or countershaft, is shown in Figs. 34-36. If from the first, there are sixteen changes, if from the second, thirty-two, independently of the speed of the spindle. The dial feed just described has been designed to supersede this.

The upper feed cone, which drives to the cones A, on the feed box is geared to the spindle, and runs at twice the spindle speed, doubling the power by comparison with a cone mounted directly on the end of the spindle. As the cones have four steps, and are interchangeable, this accounts for eight changes. By interchangeable is meant that the driving and driven cones can be made to change places, so reversing the speed relations, a smaller driving to a larger, and *vice versa;* a method which is commonly adopted in all or nearly all machines in which cones for feeds are retained. Looking at Fig. 36 it is seen that the cone fits on the sleeve which it drives by a short key over which it is slid, and it is secured endwise by the knurled nut B, so that the changing of the cones is only a work of two or three minutes. All the various speeds obtainable on each step, for the two changes of the cones, are given in the plates usually attached to machines.

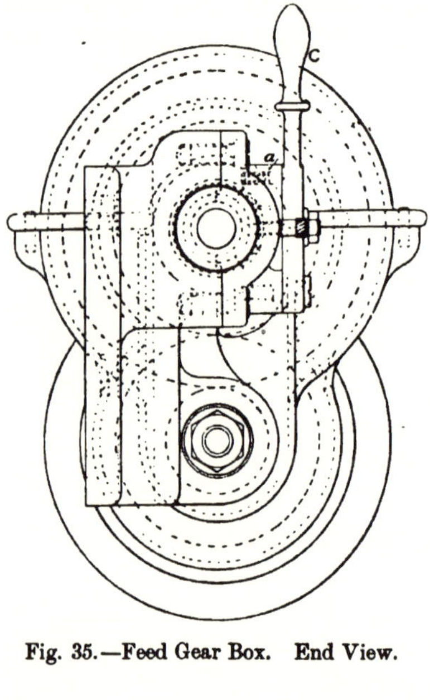

Fig. 35.—Feed Gear Box. End View.

These changes are doubled in the feed box shown by the movement of the lever C, the effect of which is shown clearly in Fig. 36. Two gears in the lower part of the box, fast on the cone sleeve, engage with two gears in the upper part of the box loose on their sleeve. Either one is caused to be driven from its lower gear by the sliding of the lever C, moving a pin and its key along into the boss of either gear. The spring pin *a*, in Fig. 35, automatically locks the lever in a position to drive one wheel or the other, or in a middle position of no drive. The great advantage of this device is that a change from a roughing to a finishing cut, and *vice versa*, can be made without shifting the belt, or stopping the machine.

The feeds transmitted from telescopic or horizontal shafts are made to actuate the two movements of the table and that of the knee in the most complete machines. In addition to these, hand movements are also fitted. Many modifications are effected in the details of these by different firms, the practice of which also changes from time to time. Reference may be made to Figs. 14-19, pages 28-32, for a present-day design.

The Knee.—The knee and its method of fitting on the latest types of pillar machines possesses several characteristic features which are worth noting. In the first place it often has a solid

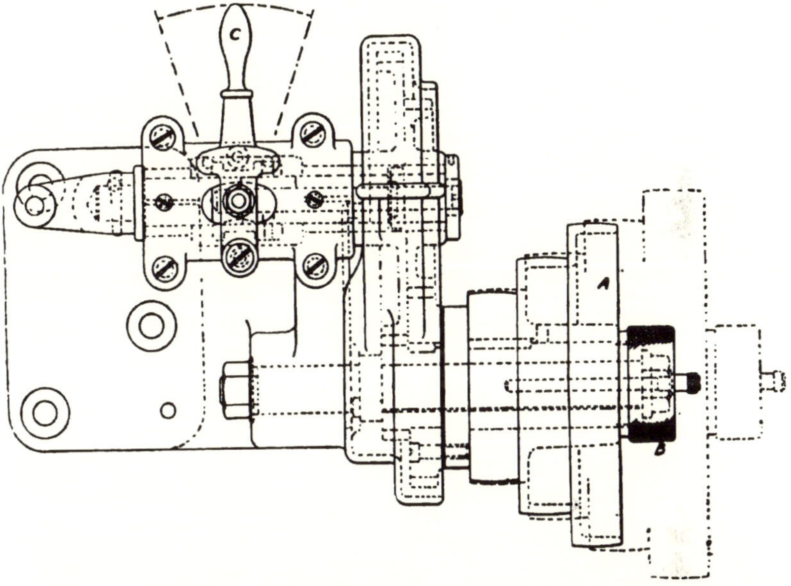

Fig. 36.—Feed Gear Box. Front Elevation. (Dotted Pulley for larger Machine.)

top, which is better calculated to resist stresses of certain kinds than one which is open on the top. The latter type is likened to a box without a bottom, which is not so well able to resist stresses as one with a solid bottom. The weakness of an open knee is generally recognised, and is to some extent lessened in many cases by the practice of casting internal ribs and fillets, so narrowing the area of the opening. Another objection to the open top is the tumbling down of chips on the screws and gears that lie within the knee. This however can be avoided by the use of telescopic slides. But the solid top at once stiffens, and affords protection. Its advantages from the point of view of stiffness are not so very pronounced when work lies within its area. But they are most evident when heavy cutting is being done with the table run out a good way over one side, with the result of producing a considerable amount of leverage, the fulcrum of which is on one edge of the knee, while the effect is to lift or attempt to lift the slide on the other edge. To prevent a wedging action against the vee'd edges under such conditions is the object of imparting square edges to the slides, but these are not considered necessary with solid knees.

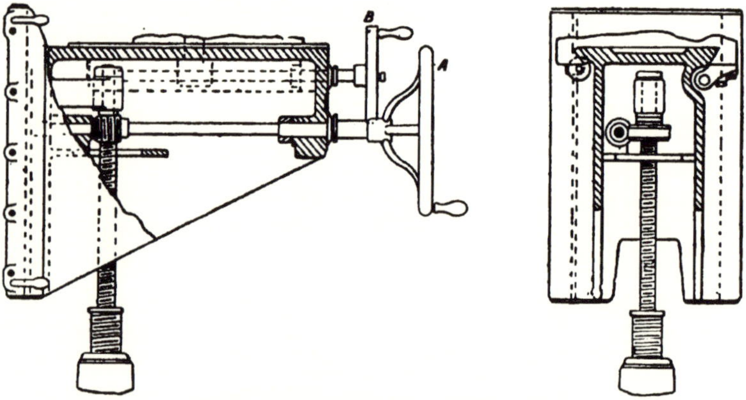

Fig. 37.– Garvin Knee, with Solid Top.

The Garvin knee shown in Fig. 37 has a solid top, and also has its sliding face extended above the top slide, a feature which is conducive to stability. In these machines the screw for the cross traverse movement is placed alongside the knee, outside to the right, lying in a recess. The nut is bolted to the slide, being also fitted thereto with a tongue. The method of taking up slackness due to wear consists in splitting the screwed boss into two portions, one of which is rigid, the other slightly elastic. The latter can be tightened against the traverse screw by an adjusting screw, effected similarly to the taking-up of the wear endwise on a worm made in two parts, by which means it is not necessary to take the precaution of allowing for backlash when making exact settings of the slides. The vertical feed screw (which is of telescopic type) is operated by a worm and worm wheel, Fig. 37, and the hand wheel for this, A; and B for transverse feed stand squarely at the front. Micrometers are fitted to both (see page 56).

Fig. 38 illustrates a great improvement on the old method of fitting the elevating screw to the knee, which involved cutting a hole in the floor. The screw is encircled with a vertically adjustable nut through which it passes, being turned either by hand or power.

Fig. 39 illustrates the Garvin stationary screw, which not only does not come through the floor, but which is fixed in the base with a Woodruff key. The rotating nut B is of good length on the screw A, and a boss extension of it comes up through the knee. The weight of the latter is taken on a ball race *a*.

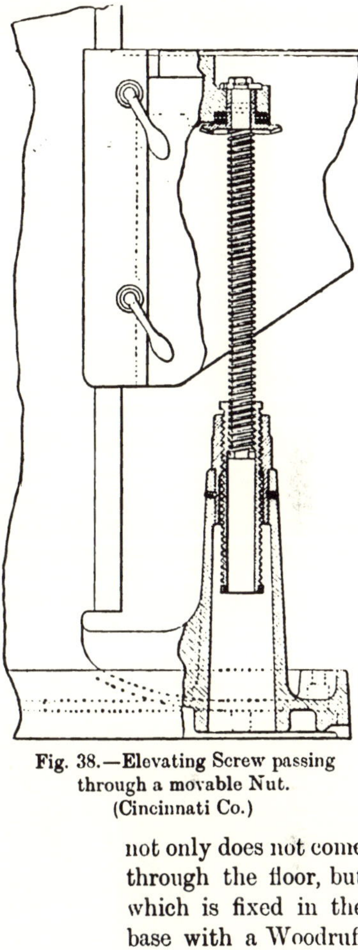

Fig. 38.—Elevating Screw passing through a movable Nut. (Cincinnati Co.)

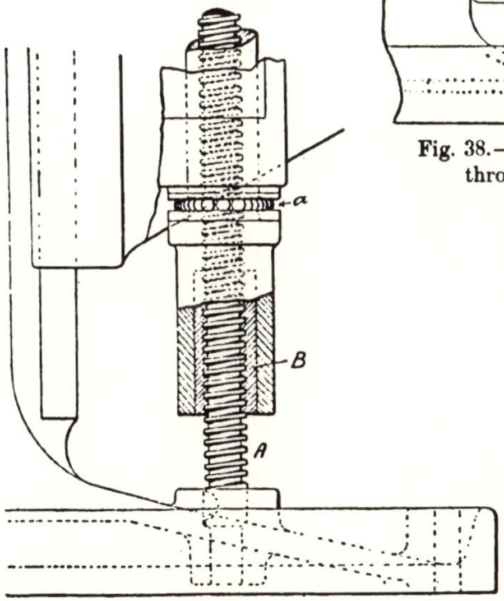

Fig. 39.—Stationary Screw.

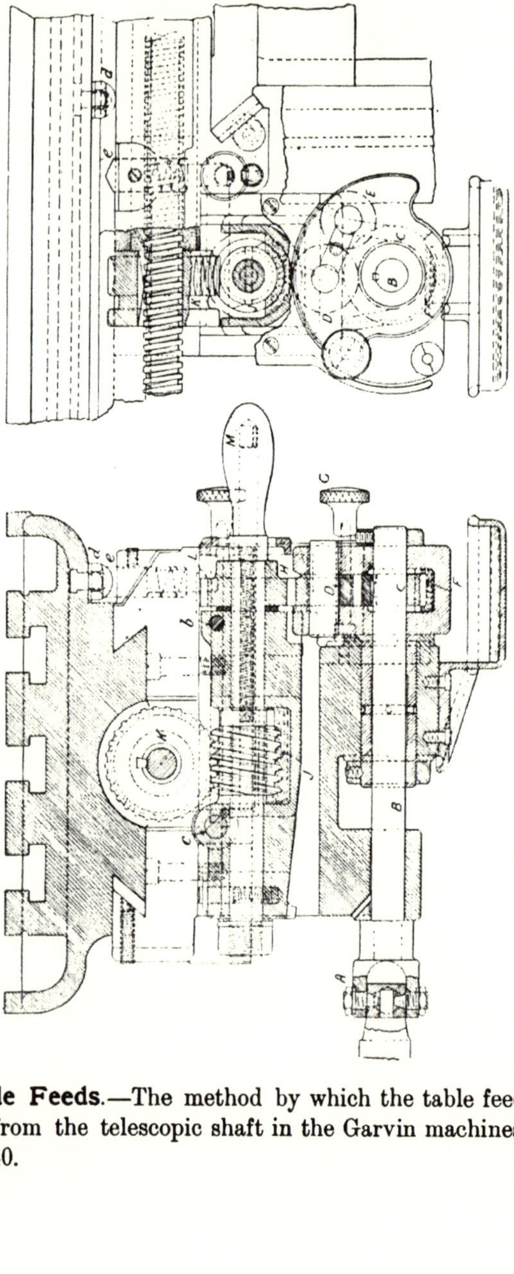

Fig. 40.—Garvin Table Feed.

Table Feeds.—The method by which the table feed is transmitted from the telescopic shaft in the Garvin machines is shown in Fig. 40.

A is the universal joint on the end of the telescopic shaft, whence the first driving shaft B is actuated. The shaft carries a gear C, that actuates tumbler gears D or E; the pins for these are carried in a swinging casting F, the bottom of which

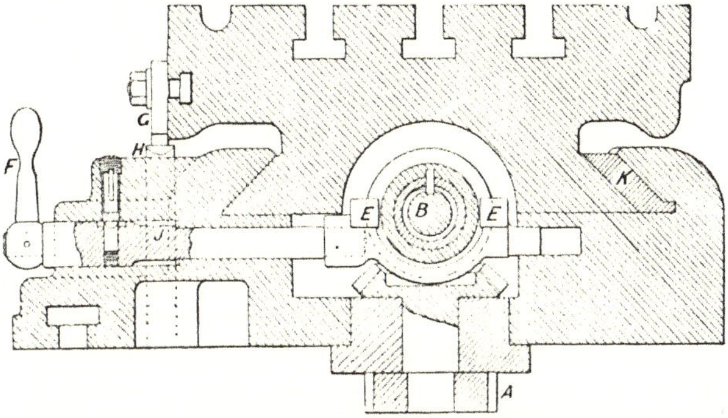

Fig. 41.—Cross Section through Table and Swivel Carriage. Le Blond Universal Machine.

forms an oil box. The reversal can be effected by hand when the machine is running by pulling out the knob G, and throwing the rocker casting F over to the other position. The feed is transmitted and tripped as follows:—The tumbler gears actuate the wheel H, on the shaft to which the worm J is fastened.

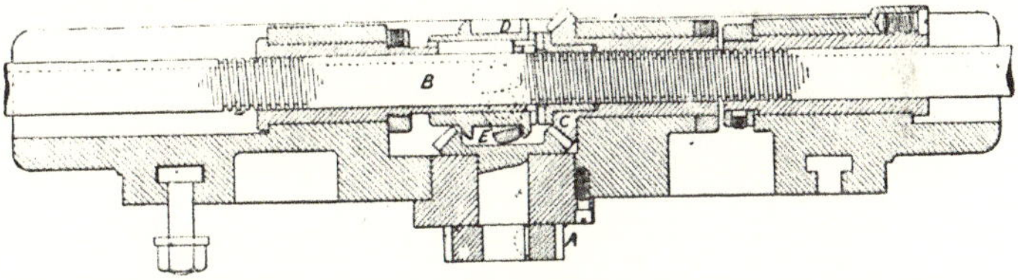

Fig. 42.—Longitudinal Section through Table Screw. Le Blond Machine.

J drives the worm K, mounted on the feed screw. The bearings of the worm shaft are in a solid casting, and include the oil bath for the worm. The casting is pivoted at *b*, and retained during feeding by the latch *c*. The feed is tripped by the button *d*, on the table, pressing down the spring plug *e*. The effect of this is to thrust the trip rod L, that supports the latch pin *c*, outwards, so letting the worm bearings drop, and disengaging the worm from gear with K. The feed can be disengaged by hand by pulling at the knob of L. It is thrown in by pressing down the handle M, which is attached to the drop bearings.

Figs. 41 and 42 show the table sections of the R. K. Le Blond Universal machine, illustrating the two methods of driving; that of a rack (not shown) from the pinion A, and that of a screw B. The feed is necessarily driven through the centre of the saddle. The screw B, adequately supported in long bearings, is

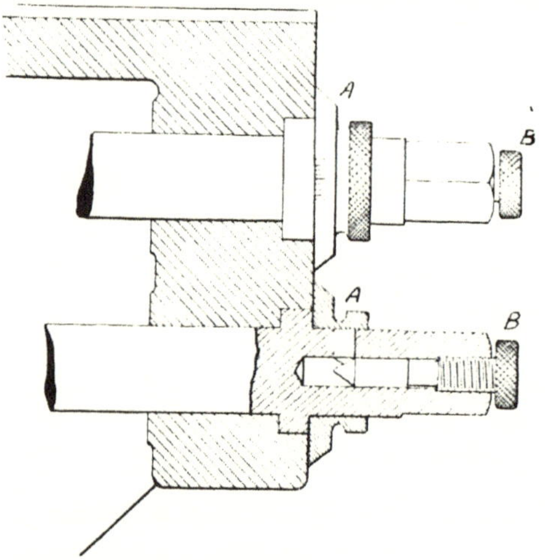

Fig. 43.—Micrometer Fittings to Feed Spindles.

encircled by a mitre wheel C, driven by one on the stem of which A is keyed. C runs freely on the screw, but is made to rotate it by the clutched sleeve D, splined to the screw and slid along by the double-forked lever E. The latter is operated by the handle F, or by the dogs, one of which is seen at G. These thrust down a plunger H, a rack on which gears with the pinion J, cut in the spindle which actuates the lever E. The gib K is a tapered one.

Fig. 43 shows the micrometer fitting applied to the various feed spindles of the Garvin machines, the upper view being an external one, the lower the same in section. On the spindle there is a steel collar A, bevelled, and graduated to indicate thousandths of an inch, and knurled to permit of turning it to start at zero. It is slackened, and again tightened when set, by the small knurled screw B at the end, which presses up a pin with a bevelled end suited to the bevelled end of the screw.

The best table stops are of precision type, set by a micrometer screw, which is bound by a clamp screw. Their value lies in the means which are afforded of feeding up to a given point, and up to shoulders.

Fig. 44 illustrates a vernier, applied to the tables of the B. & S. milling machines for the purpose of effecting fine adjustments of the table, reading to thousandths of an inch. The scale A is 24 inches long, the finer divisions being omitted in the engraving, and is attached by screws, and tee-headed nuts entering the slot which receives the trip dogs. The vernier B is attached to the front of the saddle of the machine by the bolt shown, *a* is the clamping screw for the vernier, and *b* the fine screw by which it is adjusted to zero.

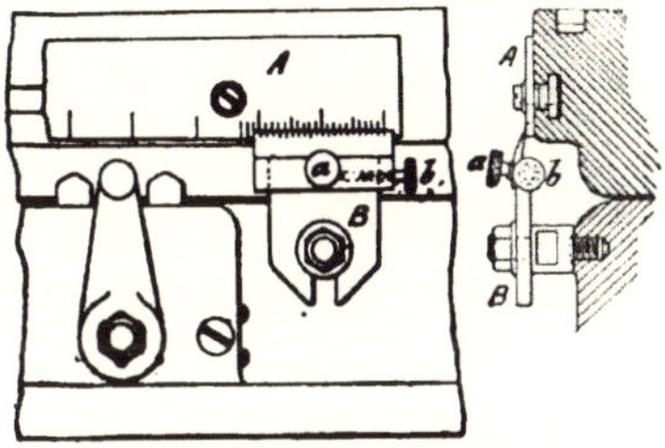

Fig. 44.—Vernier applied to Table of B. & S. Machines.

The vernier, whether applied to machine or to caliper, is a means by which fine readings can be taken by inspection from comparatively coarse dimensions. It comprises two parts, the "vernier" proper and the "beam." Different units of division are adopted by different makers, but the most useful is one which is embodied in the Brown & Sharpe vernier calipers for English divisions of a thousandth of an inch.

Figs. 45-47 show one of these much enlarged for the sake of clearness. Each inch in length of the "beam" A is divided into ten equal parts, and each of these again into four, making forty parts to the inch; so that $\frac{1}{40}$, or 0·025 inch is the value of each part. The vernier B is divided into twenty-five parts on a total length which equals twenty-four parts of the beam. These divisions are not so fine but they can be readily seen, and yet they register $\frac{1}{1000}$ inch easily, thus:—

As twenty-five parts on the vernier correspond in extreme length with twenty-four parts on the beam, = twenty-four fortieths of an inch, then $\frac{1}{25}$ of $\frac{1}{40}$ = 1,000th of an inch.

When the vernier is set at zero, Fig. 45, there is then a difference in the next two lines of division, of $\frac{1}{1000}$ of an inch, and so on,—$\frac{1}{1000}$ for each successive division, until they correspond again at the 25 division on the vernier.

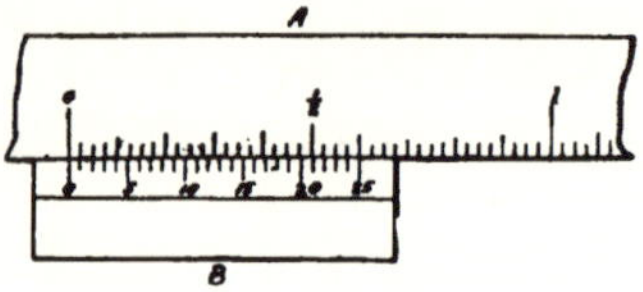

Fig. 45.—Vernier set to Zero.

If the zero marks are set together, as in a caliper, no fractional parts exist, but whole numbers, corresponding with the inches, as 1, 2, 3, &c. But when the zero on the vernier comes elsewhere, then, first, the number of inches, tenths, and parts of tenths are read, by which the zero point on the vernier has moved from the zero point on the beam. Thus in Fig. 46 it has moved from zero a distance of one tenth, and also past two of the four divisions of the second tenth on the beam. That is one tenth added to twice 0·025 inch, equalling 0·10 + 0·05 = 0·15 inch.

In Fig. 47 the vernier is shown moved past the second subdivision of the second tenth. To read this the eye must be run

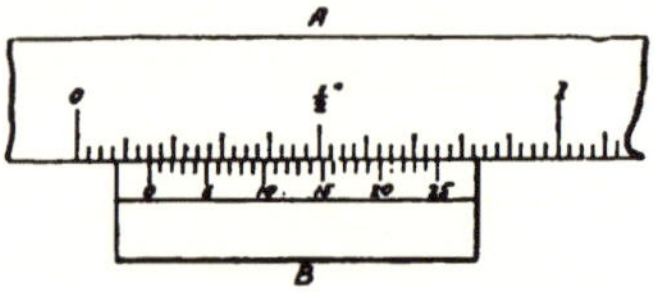

Fig. 46.—Vernier set to an Exact Division.

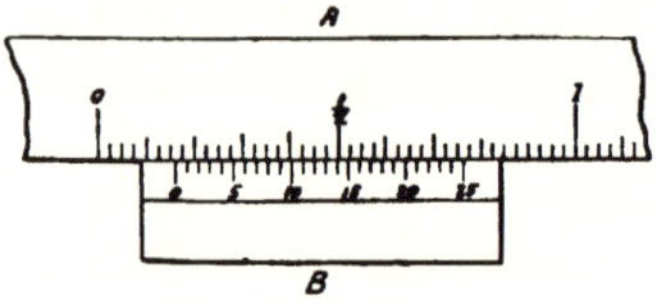

Fig. 47.—Vernier set to Fractional Division.

along to find where coincidence occurs between vernier and beam. This occurs at the twelfth division, and represents the number of thousandths to be added to the distance read off on the bar, giving 0·012 + ·15 inch = 0·162.

If the readings are taken decimally, calculation is facilitated. The tenths on the beam are then ·100, or one hundred thousandths, and the fortieths, or fourths of tenths on the vernier are 0·025, or twenty-five thousandths. Thus in Fig. 46 the jaw is open one tenth, and two twenty-five thousandths = 0·150 thousandths. In Fig. 47 it is open one tenth, two twenty-five thousandths, and twelve thousandths = 0·162.

Differences between Plain Machines and Universals.—Sometimes the plain milling machines are made with rack feeds, and sometimes with screws to the table; the universals have screws. The latter are more exact, and better suited to micrometric dimensions; the former permit of more rapid adjustments by hand. There are not so many joints for the table as in the universals, there being no swivelling arrangement; it is stiffer, therefore, which is more favourable to slogging. The universal milling machine was patented by J. R. Brown in 1865. See Fig. 5, page 17. It was exhibited at the Paris Exhibition of 1867. A modern machine by the firm is shown in Fig. 48. In this type of machine the table is capable of three movements: one, that of the table transversely along with its saddle; the other, that of the table alone, longitudinally, by means of a feed screw; the third, by the knee. The saddle swivels in a horizontal plane on a bed having vee'd edges, which bed slides along the top of the knee bracket, that moves vertically on ways on the front of the standard. The table carries the dividing head, with its gear, and the loose headstock or "footstock."

The universal machine will fulfil the functions of the plain machine of the same type. The essential difference between a plain machine and a universal is, that the latter has, in addition to the fittings of the former, a spiral head, index plate, sector, and change gears, with a swivel table. The index plate furnishes circles of divisions, change gears give variable rates of spiral movements, while the table is being traversed a certain distance by the feed screw. By these additions any angle up to 45° can be given to the table, which can also be fed automatically at any angle, and a definite rotary movement can be given to the work. There is nothing therefore in the range of small gear cutting, cutter or tool making, or the tooling of small parts within the dimensions covered by the machine, which cannot be accomplished on the universal type. It owes its chief value therefore to the fact that there is no job of milling of moderate dimensions which cannot be done upon it. Every shop should have at least one of these, but it is a mistake to suppose that it is economically adapted for all kinds of jobs, even though these may be of medium dimen-

Fig. 48.—Universal Milling Machine. (Brown & Sharpe Manufacturing Co.)

sions. Many such can frequently be tooled to better advantage in large quantities on the larger and more specialised machines of various kinds. It is not judicious to attempt heavy tooling on a universal, for which almost any other types are better suited. It is a somewhat delicately-constructed piece of mechanism by comparison with some others. The table is not quite so rigid under stress; and as it is driven by screw instead of by rack, the truth of the screw should be retained as long as possible. It is a little top-heavy. The universal head is a piece of delicate construction; there is a deal of gear introduced for the automatic movements, all of which details are incompatible with heavy tooling. Only work should therefore be put on the universal which cannot be done at all, or at least so conveniently elsewhere, and specially that which requires the assistance of the head and the swivelling arrangements of the table. A skilful man is required to attend to a universal machine, because the work done upon it is of an intricate character, and also because calculations have to be made for the cutting of teeth of spiral form; and when gears are cut, calculations also for the numbers of teeth. A good attendant will turn out a large volume of excellent and varied work from a universal machine, and he will prove the best investment.

The table is fed automatically from the cone pulleys through a telescopic rod and worm gear in the older machines, but through bevel gears and a screw in others; and the automatic movement can be stopped or reversed. The automatic drive is effective at any angle of the table. The table can be operated by hand from either end. The movements of the table in any direction, including the angle to which it may be set, are indicated by micrometer dials, reading to thousandths of an inch. Changes of feed range between eight and thirty-two in number on different machines, the wide range being obtained by transposing stepped cone pulleys and pulleys at the back of the machine, or by gears, or both in combination.

The handles for operating the three movements of the table—longitudinal, transverse, and vertical—are at the front; and two of these—that for the cross feed and that for the vertical movement—are often provided with clutches. The idea is that the handles can be instantly disconnected; and being loose on their respective shafts, the adjustment of the slides would not be affected should the attendant accidentally shift the handles round. At the same time, they are ready for immediate use.

The handle for the vertical movement of the table by hand is often placed at an angle, instead of just below the cross-feed handle, so that they do not interfere with one another, neither has either of the handles to be removed, but remain permanently on their spindles. It is easy therefore to operate two handles at one time.

Ball bearings are commonly applied to the feed screw of the table, as well as to the vertical feed screw. Quick return can be employed to the table. Its base is of large diameter, and graduated on the outside into degrees for milling spirals, or work of similar character.

In conformity with an agreement adopted by the manufacturers of milling machines in the United States the movements on plain milling machines are as follows :—

	No. of Machine.						
	0	1	1½	2	3	4	5
	Inches.	Inches.	Inches.	Inches.	Inches.	Inches.	Inches.
Transverse movement	6	7	7	8	10	12	12
Vertical movement -	15	19	19	19	20	20	21
Automatic table feeds	18	24	24	28	34	42	50

The movements on universal milling machines are as follows :—

	No. of Machine.			
	1	1½	2	3
	Inches.	Inches.	Inches.	Inches.
Transverse movement - -	7	7	8	10
Vertical movement - - -	18	18	18	19
Automatic table feeds - -	20	20	25	30

Index Centres and Spiral Heads.—The index centres used on milling machines range from those which are quite plain to those which are universal in character. A plain type, of which there are several varieties, consists of centres without any provision for setting to a vertical angle, and without any means of sub-division in the head except that afforded by holes drilled in a dial. A tapered pin inserted in the holes locks the spindle while teeth are being cut or edges milled equidistantly.

In a more elaborate kind, the centres, as before, have no means

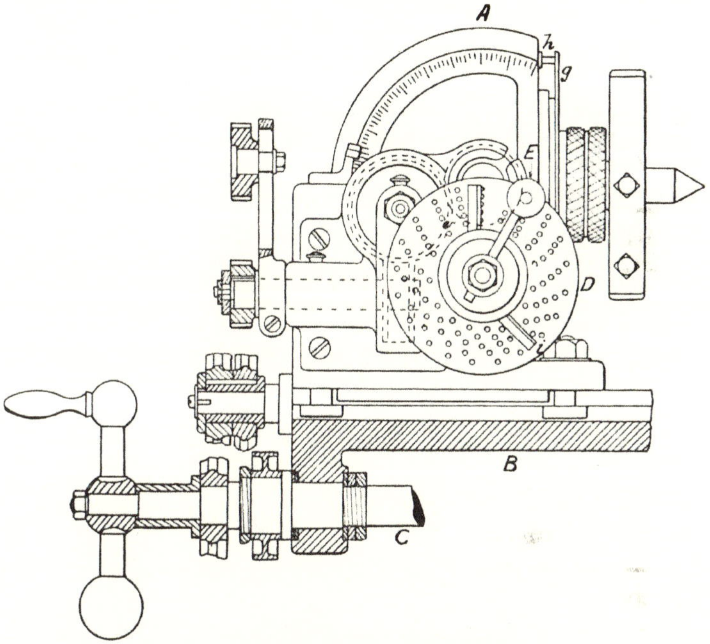

Fig. 49.—Universal Indexing Head.

of angular adjustment, but a nearly universal range of indexing can be obtained by the employment of index plates with circles of holes and a worm and worm wheel, the latter being on the headstock spindle.

In the most perfect type the index head will angle from a few degrees beneath the horizontal to a few degrees over the perpendicular, giving a total range of 118° or 120° of adjustment in a vertical plane. Index plates in combination with worm gears give a large range of numbers for division. The spindle also of the loose headstock has a slight vertical range of angular adjustment, suitable for taper taps and reamers and work of that type, and a vertical movement also bodily. Such centres fulfil universal functions, including the cutting of all gear wheels and spiral cutters and spirals. The universal head, in this its most perfect form, combines two movements, one of rotation and division in itself, the other one, of longitudinal movement through the table; the first being effected through a division plate, the second through change gears and the table feed screw. By the first it is possible to mill out equal divisions, as on gear wheels and cutters; by a combination of the two, any number of spirals situated equidistantly can be milled. The feed screw, being of a definite pitch like a lathe feed screw, is the basis for the calculation of change gears. The division plates used are like those on geometric lathes, divided into circles of holes which receive the index peg.

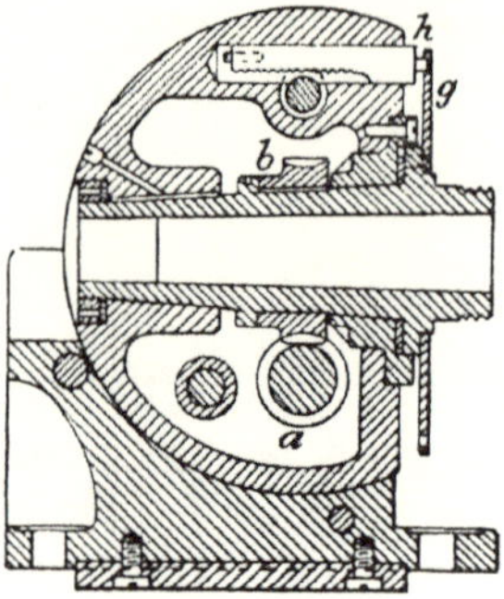

Fig. 50.—Universal Head. Vertical Section.

Figs. 49-52 illustrate one of the dividing heads of this class as fitted to the Nos. 2 to 4 Brown & Sharpe universal machines. Fig. 49 is a view taken partly outside, partly in section, showing the head A fitting on the table B of the machine, C being the feed screw. The head can be revolved either by hand or by the feed screw. Enclosed within the head A are a worm and worm wheel, seen in Figs. 50 and 51 at *a* and *b* respectively. The

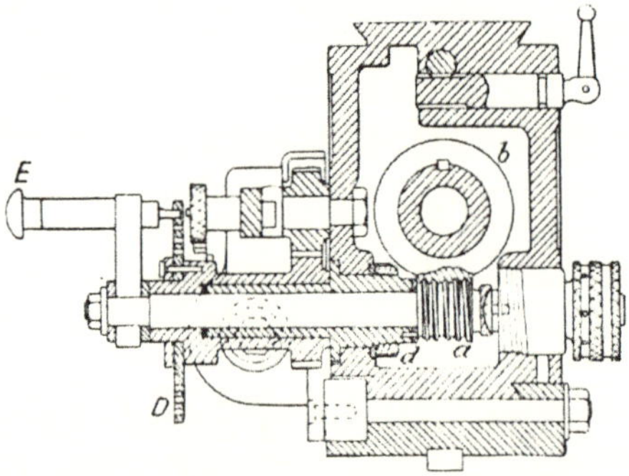

Fig. 51.—Universal Head. Transverse Section.

latter rotates the spindle, irrespective of any angle at which the spindle may be set. The worm *a* is turned by the crank handle E, Fig. 49, or it is rotated automatically by the feed screw C through the change gears at the left. These operate through spiral wheels enclosed in the head, and seen dotted in Fig. 49. The worm gives divisions; the change gears impart precise rotary movement to the spindle during cutting, which produces spirals. The spindle can also be rotated by hand when cutting coarse divisions like those required in reamers, end mills, &c. The mechanism by which this is accomplished is seen in Figs. 51 and 52. It is necessary first to throw the worm out of gear with its wheel, which is done as follows:—The knob *c* is turned by a tommy about a quarter of a revolution in the opposite direction to that indicated by an arrow which is stamped on the knob *c*. This loosens the nut *d*, which retains an eccentric bushing *e*. Then with the

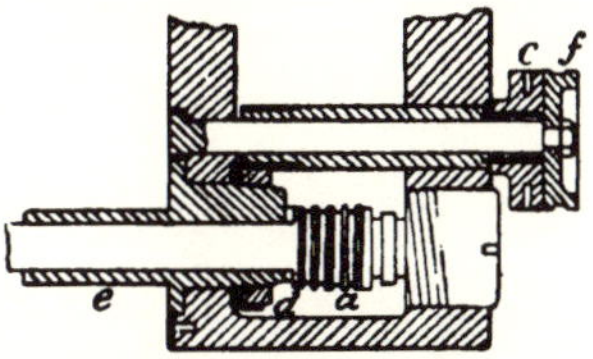

Fig. 52.—Universal Head. Disengaging Motion.

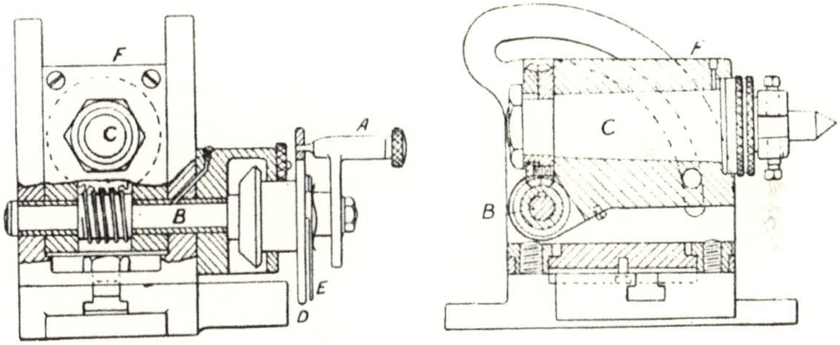

Fig. 53.—Universal Head.

fingers the knobs *c* and *f* are turned in unison, when the eccentric bushing *e* will be revolved, and the worm *a* thrown out of gear. The spindle can then be turned quickly by hand, and locked by the plate *g* and pin *h*, Figs. 49 and 50. To throw the worm into gear again, turn the knobs *c* and *f* in the direction which the arrow indicates, and tighten the knob *c* with the tommy.

In these machines the cutting of 68 different spirals is provided for. Three index plates are sent with the machines, having 15, 16, 17, 18, 19, 20; 21, 23, 27, 29, 31, 33; and 37, 39, 41, 43, 47, 49 holes respectively. The plate *g*, for rapid indexing on the nose of the spindle, has 24 holes.

Fig. 53 shows the spiral head used on the No. 1 B. & S. universal machines. It is simpler than that just illustrated.

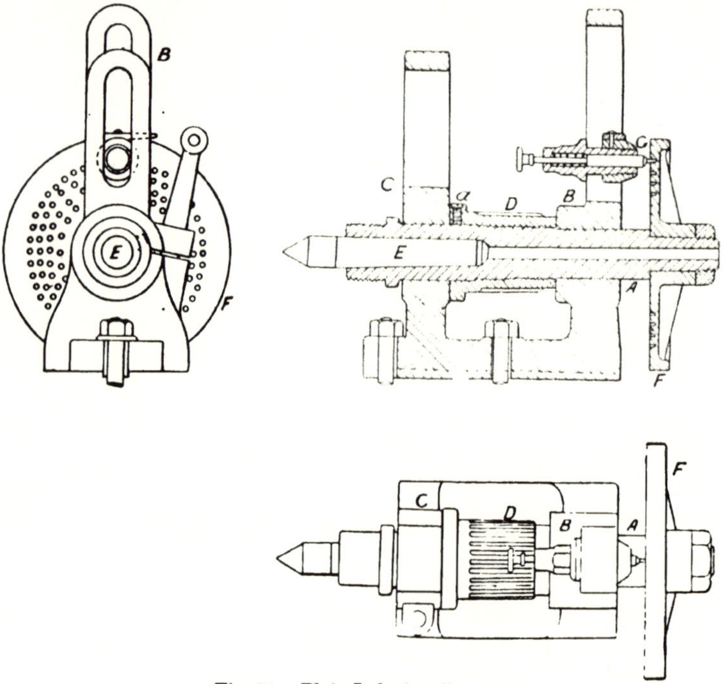

Fig. 54.—Plain Indexing Centres.

In this, as in the former one, the worm wheel has forty teeth, so that one turn of the index crank A, and worm shaft B, moves the spindle C through one-fortieth of a revolution. By means of the index plate D, the fortieth of a revolution can be further sub-divided, and the sector E is set to divide the holes without counting. The crank A is adjustable radially, so that the pin can be used in any circle of holes.

Bushings of steel encircle the spindle B and form a durable pivot for the box F. The spindle and box can be swivelled, and set to any angle from 5° below the horizontal to the perpendicular. One side of the head is graduated. Change gears transmit motion from the feed screw to two mitre gears, one of which is seen in the Fig.

Plain indexing centres by the Cincinnati Milling Machine Company embody some special features. The headstock has no power of angular adjustment; but instead, the centre of the

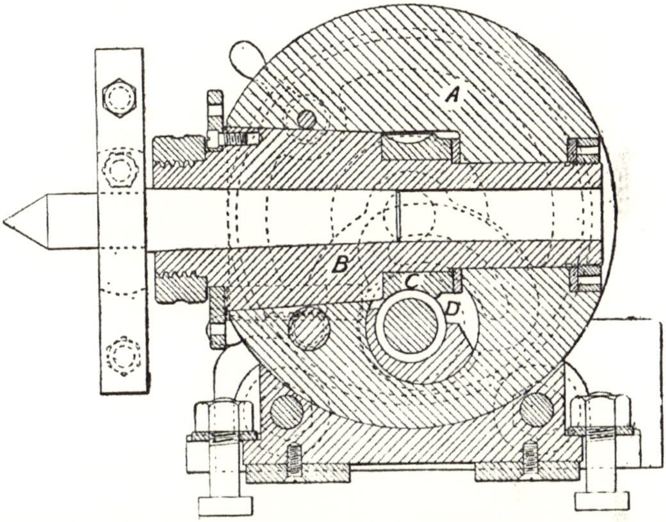

Fig. 55.—Le Blond Dividing Head. Longitudinal Section.

tailstock is made movable in a vertical direction. The principal elements of these centres are shown in Fig. 54, in sectional elevation, plan, and front-end elevation.

The spindle A is carried in bearings B and C, the latter of which is split to clamp the spindle, this being rendered necessary by the fact that it has a power of endlong movement imparted to it, because there is no such motion in the spindle of the tailstock. To move the spindle endwise, grooves are formed for the grip of the hand on the body of a sleeve D, which encircles the spindle A about the centre, the spindle and sleeve being mutually threaded.

The sleeve on being turned by hand, while coerced by the uprights B and C, moves the spindle A endwise. In order that the sleeve shall not move accidentally, and that it shall turn when the index plate is moved, it has frictional connection with the spindle by means of a small screw *a* pressed on the spindle. By means of the split bearing C the spindle is clamped after adjustment.

The spindle has a tapered socket at the front end to receive the centre E, while a thoroughfare hole permits of the insertion of a bar for shooting the centre out.

The indexing is done by the plate F at the rear of the spindle, two or more of which are supplied. The pin G is adjusted up and

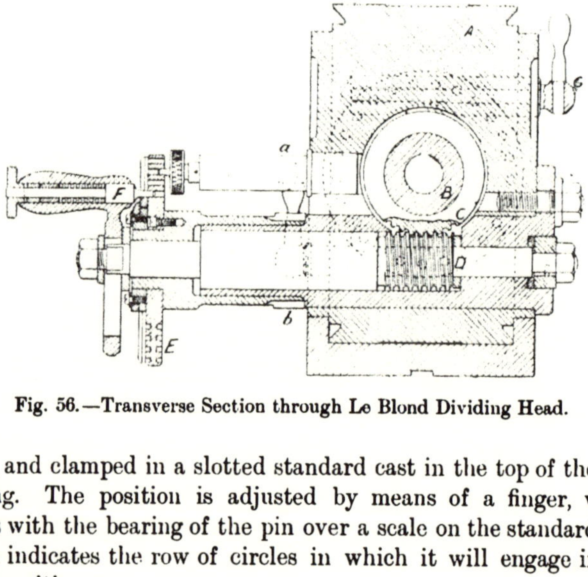

Fig. 56.—Transverse Section through Le Blond Dividing Head.

down, and clamped in a slotted standard cast in the top of the rear bearing. The position is adjusted by means of a finger, which moves with the bearing of the pin over a scale on the standard, and which indicates the row of circles in which it will engage in any given position.

Figs. 55 and 56 illustrate the Le Blond dividing head, the leading features of which are the following:—The body is circular in form, and a solid casting, capable of swivelling through an arc of 200°, or 10° below the horizontal on each side. A dovetail is turned on each side, Fig. 56, by which it is clamped to the base with two bolts, the heads of which are turned to the radius of the dovetail. The drive from the table screw, through simple or compound gear, takes place through a bracket bolted to the side of the base (not shown), and which carries the quadrant for the change gears.

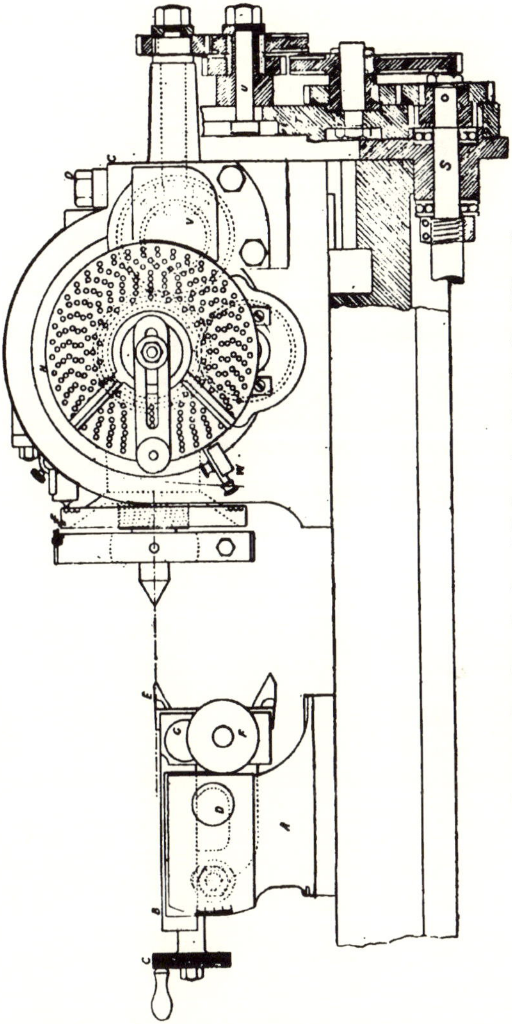

Fig. 57.—Dividing Head, and Tailstock. Longitudinal Elevation.

At the opposite end to the quadrant, a bevel wheel engages with a bevel (not shown), running on a stud *a*, Fig. 56, in the centre of the swivel head, and a spur gear connected to this bevel drives a spur *b* on the sleeve of the index plate. The positions of the gears are thus maintained at all angles of the swivel head. The head A is bored taper to receive the spindle B. A worm wheel C encircles the spindle, rotated by the worm D, solid with the spindle that carries the division plate E; F is the index pin. The spindle is clamped by the locking plug G. The worm and its shaft are carried in an eccentric sleeve provided with a slot, and stop, so that it can be thrown into and out of gear rapidly. The worm gear can thus be quickly disconnected when quick indexing is required by a

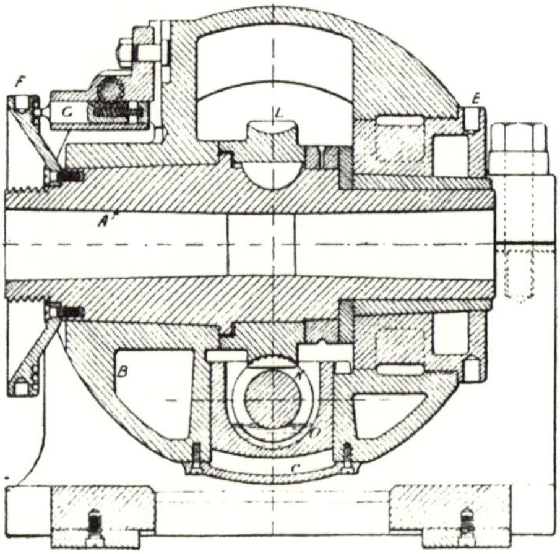

Fig. 58.—Longitudinal Section of Dividing Head.

dividing plate alone, as in fluting taps, reamers, and cutting some small gears, sprockets, &c.

Figs. 57-59 illustrate a spiral headstock by the Cincinnati Milling Machine Co. Fig. 57 shows the head in elevation. Fig. 58 is a longitudinal section through the head, and Fig. 59 a transverse one.

Heads are mostly confined to angular movements, ranging from 10° below the horizontal, to 10° beyond the vertical. In this example the spindle A is carried in a swivelling block B, which is capable of a complete revolution in a vertical axis, and the edge of which is graduated into degrees. An advantage of this complete swivel is that it permits of the cutting of right and left hand work without changing the cutter, and the cutting of the same piece on each side of the vertical. Another is that the swivelling block is always in any position contained wholly within its bearings, and therefore adequately supported.

No strain is thrown on the spindle in clamping, which is done on two clamping rings C, C, Fig. 59, embracing trunnions on the block. The rings are pinched by the cap screws D, Fig. 57. As the circumference of these is so large, no strain or distortion can affect the spindle, or the worm gear.

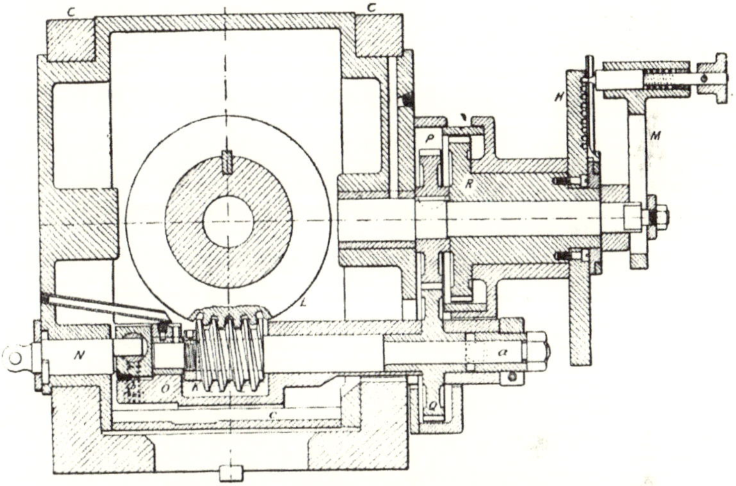

Fig. 59.—Transverse Section of Dividing Head.

The spindle A is provided with a clamping ring at E, which secures it endwise. Wear is taken up between the coned neck at the front of the bearing, and the coned bushing at the back.

The dividing mechanism includes the worm and wheel, index plate, and sector, and another index plate F, added for low numbers, or those under forty, the ordinary sector plate being reserved for those over forty.

Plate F, used for direct indexing, is provided with three circles of holes on the back, namely, 24, 30, and 36, into which the index pin G is inserted. The holes for divisions 4 and 6, which are so often wanted for fluting and nut milling, are indicated by corresponding figures on the edge to save the trouble of counting round.

The regular index plate is seen at H with its index pin. It is drilled from both sides to increase the divisions by simple reversal of the plate. Divisions are obtained by this in conjunction with the worm K and its wheel L. The worm—right handed—is single threaded, with 2½ threads per inch. The wheel has forty teeth, so that forty turns of the index handle M turn it through one revolution. The handle M turns the worm and wheel through equal gears, P, Q, for dividing, or indexing only. For spiral work, R is the gear through which changes in pitch are effected through trains of wheels.

The worm K can be thrown out of gear by dropping it, with its bearings, bodily, by the eccentric pin N, the case being hinged about the pin *a* as a centre. As the casing O is confined in the lateral direction, the eccentric end of the pin is inserted in a bushing, which has an endlong movement in the worm case. The latter is fastened to the block containing the bushing with screws, which afford means for adjusting the worm to the wheel. The worm runs in a bath of oil, and this can be cleaned out on removal of the cap *c*.

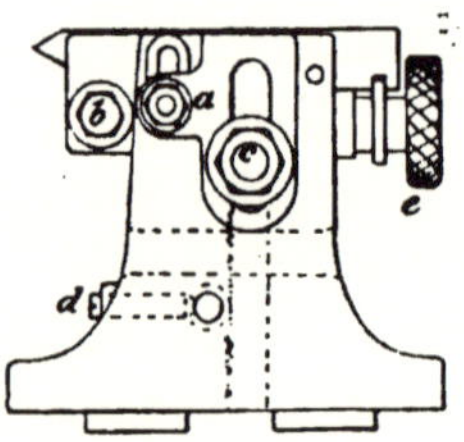

Fig 60.—Footstock.

The change gears are seen in Fig. 57, S being the lead screw, with ball thrusts, T the swing plate, and U one of the two studs for compound gears. Motion is transmitted through the bevel wheels V actuating the spur wheels in Fig. 57.

At W a ratchet and pin are shown which permit of making fine adjustments of the index plate for the purpose of re-setting work.

Footstocks.—The footstock of the Cincinnati Company is shown in Fig. 57. The housing A carries the longitudinal slide B, fitting with a dovetail. It is traversed by the knob C, and clamped by the bolt D. The centre E is double ended, one end being the ordinary centre for heavy work, the other being reduced on the top face to permit cutters to clear it when depthing flutes. The centre is therefore reversible. It is adjusted vertically with a rack and pinion, actuated by a knob F, and is clamped by the bolt G. The centre can be tilted forwards, or swivelled about the centre D, exact angles being given by graduations at the rear of the housing.

The Brown & Sharpe footstock is shown in Fig. 60. Its

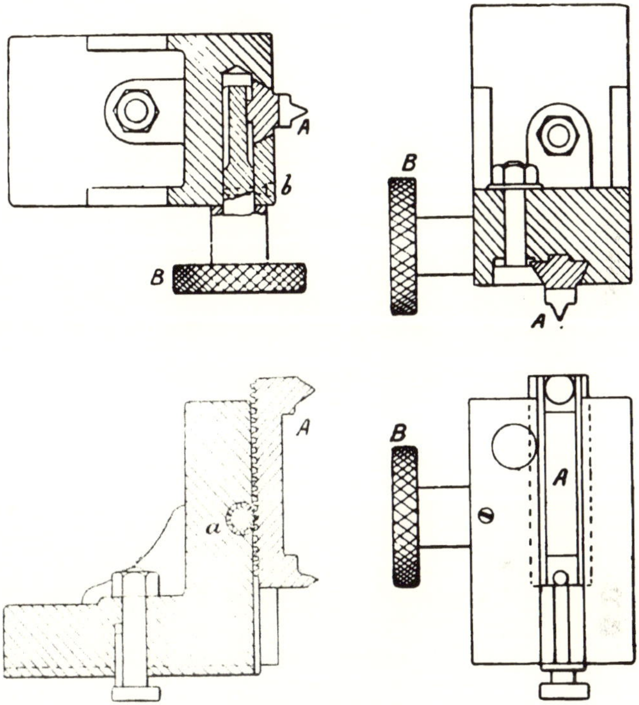

Fig. 61.—Footstock.

centre is adjustable for milling tapered work. Being adjustable, it is set in a horizontal position by taper pins. It is elevated and depressed by means of a small rack and pinion through a nut. The nuts *a*, *b*, *c* clamp the centre in position.

Fig. 61 illustrates another tailstock in two horizontal sections, in vertical longitudinal section, and in plan. The bar centre A, it will be seen, moves vertically in the body in a dovetailed groove. It has rack teeth on the back, which are actuated by a long pinion *a*, turned by a milled head B. The bar A has two centres: one for stiff work of large diameter, the other for that of small diameter, the face being cut away close down to the centre. The pinion spindle is prevented from endlong movement by a set screw *b* entering into a turned groove. The centre A is clamped by means of a bolt, the head of which is chamfered to bed against the bevelled edge of the sliding centre.

CHAPTER III.

ATTACHMENTS AND BRACINGS.

Vertical and Angular Spindle Attachments to Horizontal Machines—Examples—Heads that Swivel Bodily—Examples—Slotting Attachments—Examples—The Overhanging Arm—Bracings—Examples—The Work of the Pillar and Knee Machine.

Vertical and Angular Attachments to Horizontal Machines.—Several firms make an attachment for their universal milling machines by means of which the direction of cutting can be changed either to the vertical or to any angle. It comprises a tool-holder which is clamped to the overhanging arm vertically or at an angle. Its main spindle is driven by the horizontal spindle of the machine, and drives through spiral gear or through bevel gears the cutter spindle. It renders the machine available for revolving cutters for key seating, for tee slots, for sawing off work to definite lengths, &c. The spindle can be set at any angle in a vertical plane through a graduated index.

The convenience of being able to do vertical, horizontal, or angular milling on one machine is largely a concession to the small shops. None of the early machines had this convenience. They were either horizontal or vertical. Now there are many machines of horizontal design that are rendered capable of vertical milling, and many vertical spindle machines with provision for horizontal milling. Angular work is usually included. These are often effected by separate attachments, rather than by the swivelling of the main spindle bearings; but there are many instances of the latter, especially in Continental practice, where it is very common, examples of which are given on pages 78-82.

Fig. 62 illustrates a vertical spindle head by the Cincinnati Company attached to a horizontal machine. It is fixed to the piece A bolted to the headstock, by a large circular flange B, 8$\frac{5}{8}$ inches in diameter, and four $\frac{5}{8}$-inch bolts in an annular tee groove C, by which a movement of 360° is available. A subsidiary advantage of this swivel is that the head can be swung to an angle for the purpose of changing cutters conveniently, which is done more easily thus

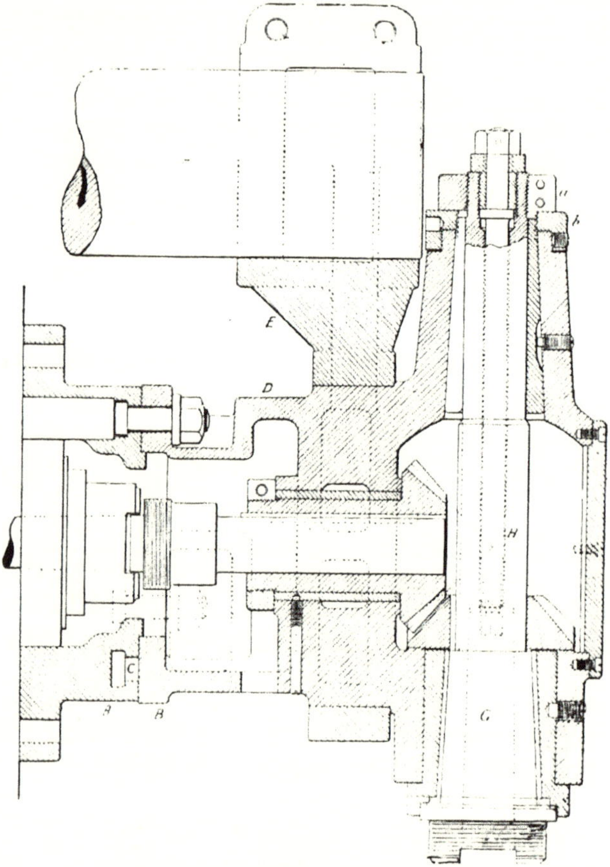

Fig. 62.—Vertical Spindle Attachment.

than with a rigidly fixed vertical spindle. Further support is afforded to the swivelling body D by a bracket E, clamped to the overhanging arm F.

The spindle G is tapered at the lower end for take-up of wear by the lock-nut *a*, and runs in a bearing of babbitt. The upper bearing is of bronze, tapered externally for taking up its wear, the taper being drawn along in the bore of the head by the nut *b*. The lower end of the spindle is clutched to match clutches on the

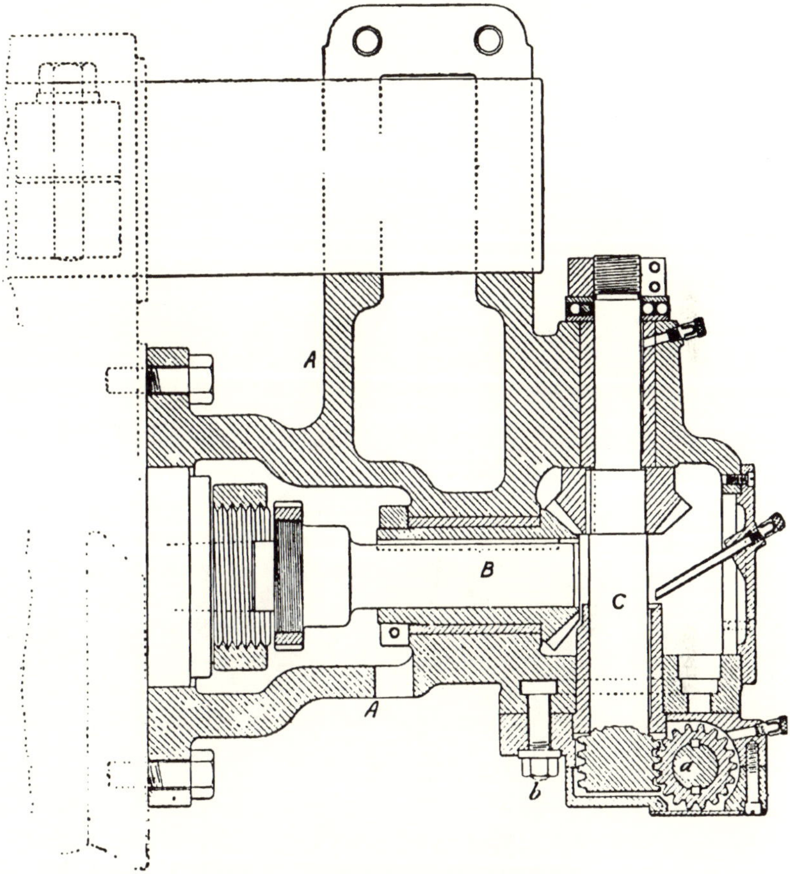

Fig. 63.—Attachment for Spiral Milling.

shanks of the cutters, which affords a positive, or non-slipping drive. The taper is the same as that of the horizontal spindle, so that arbors are interchangeable on both. The ends of the shanks are tapped to fit the $\frac{5}{8}$-inch bolt shown at H, by which they are secured. The shoulder seen next the tail of the bolt serves for backing a shank out of its hole, as well as for drawing it in. The nose of the spindle is threaded externally, like the end of the horizontal spindle of the machine, so that chucks or disc cutters are interchangeable on both.

The illustration, Fig. 63, is that of an attachment made to the Cincinnati milling machines, to permit of doing spiral milling when the angles exceed 45°. It will mill these when either right or left handed, and also plain spur gears and racks.

The casing A of the attachment is bolted to the front of the headstock, and further steadied by the overhanging arm. The cutter arbor *a* is slewed to any horizontal angle, and set by the bolt *b* by the graduated horizontal edge of the head A. The drive takes place from the main spindle through an arbor B, which fits the spindle. It is feather-keyed into the driving bevel wheel, which rotates the vertical spindle C through another bevel wheel. From C the cutter spindle *a* is driven through spiral gears, having a ratio of 2 to 1. The advantage of this is, that as the speed of the cutter arbor is only half that of the main spindle, the belt is run at twice its usual speed, and power is gained for heavy milling. Gears of 3 diametrical pitch are thus milled at a single cut. The thrusts of the vertical and cutter spindles are taken with ball bearings. A gauge is provided for setting the cutter centrally with the blank. It can be adjusted laterally to bring the vee central with the tooth of the cutter, and then reversed to bring the centre of the tailstock central with the vee.

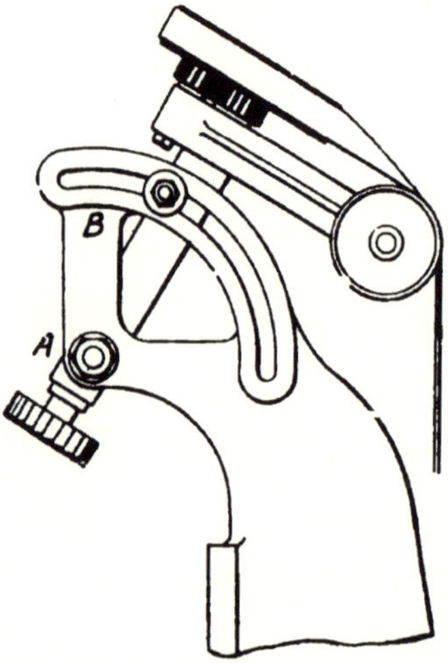

Fig. 64.—Swivelling Spindle Bearing.

Swivelling Heads.—A large group of machines have heads

that swivel bodily, in preference to making loose swivelling attachments to ordinary machines. Arrangements are very diverse. A selection, therefore, is given here.

The firm of Prétot, of Paris, make machines in which the spindle bearing is pivoted near the cutter end, being moved over a quadrant B at the other, Fig. 64, in which it can be clamped at any angle from vertical to horizontal. Adjustment is made by hand in the smaller machines, but in the large ones by a hand wheel, which actuates mitre wheels that turn a screw in a nut, which is connected

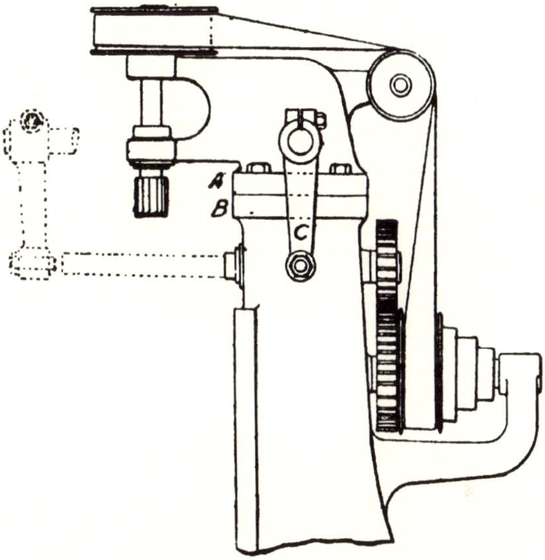

Fig. 65.—Swivelling Head.

to the bearing through the slot of the quadrant. The screw is pivoted to give freedom of movement at varying angles. The driving and the guide pulleys are carried in a cast-iron frame of U shape, which is attached to the upper end of the spindle bearing. The spur gears, seen within the belt pulley, are used in conjunction with two others, hidden by the pulley, as back gears, giving a slow drive, more powerful than the belt alone will carry.

Vautier & Co. obtain vertical and horizontal movements by using two separate spindles, Fig 65. The vertical one has to be swivelled to one side to permit of the horizontal one being used. To permit of this, the vertical head A swivels on top of the pillar B to any horizontal angle, a provision which permits of doing angular milling without swivelling the table. The horizontal cutter arbor is carried between the horizontal spindle and an overhanging bracket C having a centre, which arrangement is indicated by dotted lines. The spindles are both belt-driven, the vertical one direct over guide pulleys, the horizontal one from the cone pulley through a pair of equal spur gears.

Fig. 66 illustrates the arrangement in the (French) Bariquand & Marre machines. The head A can be set to any vertical angle

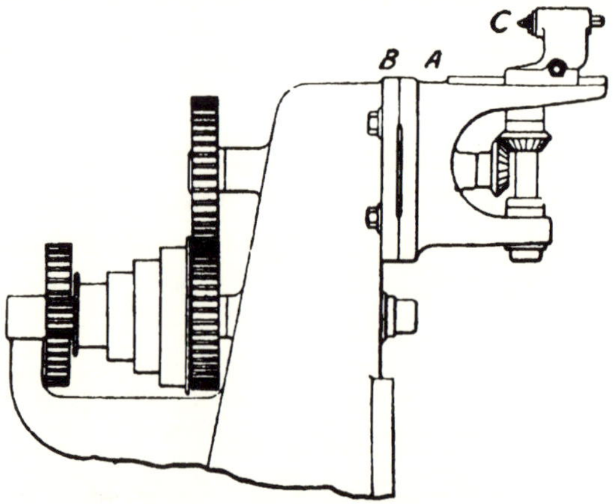

Fig. 66.—Swivelling Head.

on the face B of the pillar. The vertical spindle is driven from the horizontal one, through spur and mitre gears. When the horizontal is in use, the head is turned round 180° on the vertical face B, to bring the overhanging centre C for the outer arbor support into line with the horizontal spindle.

A head by Sculfort & Fockedey has the spindles arranged as in Fig. 67. The horizontal spindle is driven directly by the stepped cones, or through back gear. That of the vertical is derived from the horizontal, through the long pinion A, sliding on its shaft, and pushed into engagement with the gear C by the knob B. A is always in gear with a pinion enclosed in the hood D, on a

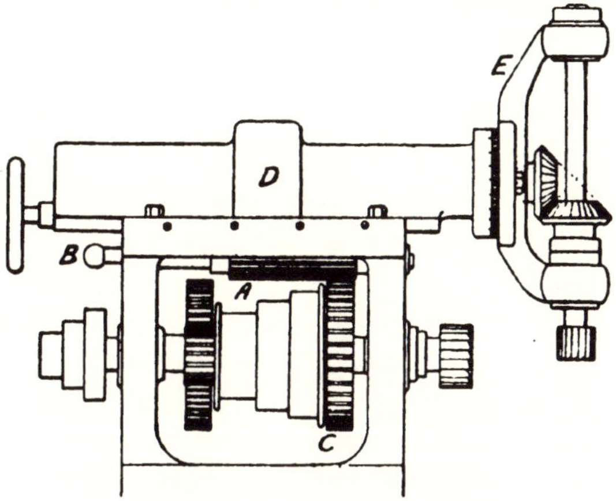

Fig. 67.—Swivelling Spindle Head.

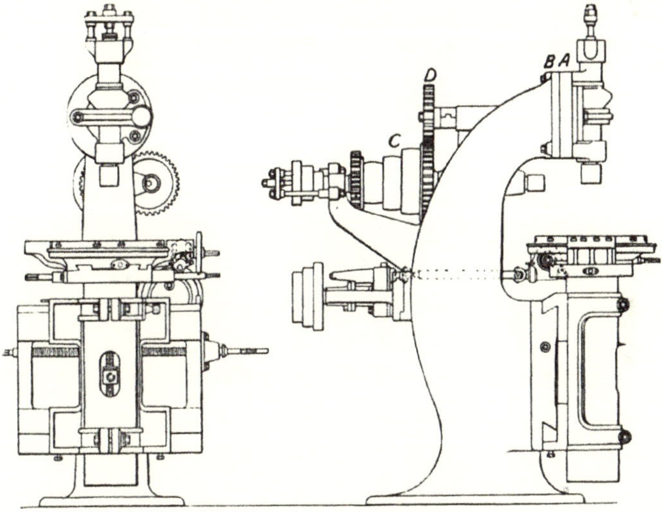

Fig. 68.—Swivel Head Machine.

horizontal shaft, which drives the vertical spindle through mitre wheels. The bracket E can be set to any vertical angle on the face of the graduated arm.

Figs. 68 and 69 illustrate a machine by E. Dubosc, of Turin.

The head A is adjustable for vertical angle on the face B of the pillar, to bring either one of the two spindles into operation. The horizontal spindle is driven directly through the stepped cones C, with or without back gears. The vertical spindle is driven from the spur wheel next the large cone step through a spur wheel D. The swivelling head has a socket to receive a steady bar E, which carries a centre to support the overhanging end of the horizontal arbor.

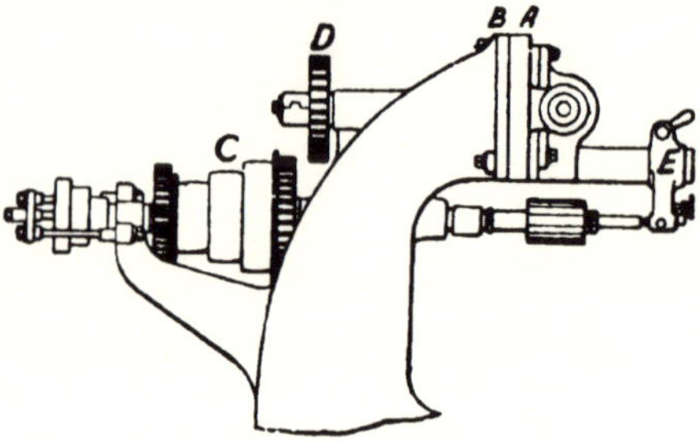

Fig. 69.—Head of Dubosc Machine swivelled for Horizontal Work.

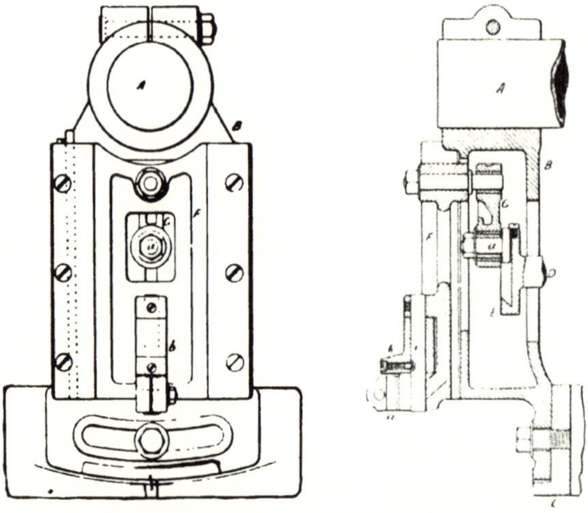

Fig. 70.—Slotting Attachment for Pillar Machines.

Slotting Attachments.—Fig. 70 illustrates a slotting attachment fitted to the Brown & Sharpe pillar machines. The overhanging arm A receives the main casting B, clamped thereon by the boss and split lug above. B is provided with a curved slot below, which permits it to be radiated about the arm A, within the limits of 10° on each side of the perpendicular, so that tapered edges can be tooled as well as those which stand square, which provision renders it suitable for the cutting of smiths' stamping dies. To increase its rigidity, the main casting is bolted through the curved slot to a piece C, which is attached to the slide on the front of the column of the machine.

The ram is driven in the first place from a tapered shank,

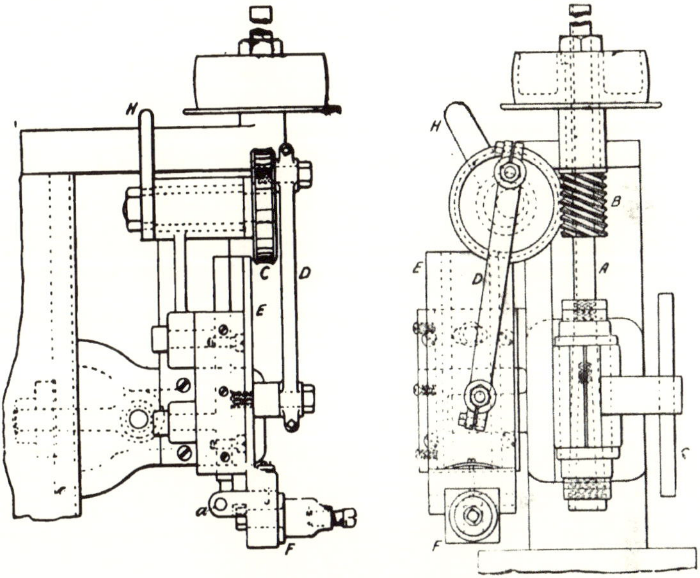

Fig. 71.—Slotting Attachment.

shown broken at D, that fits into the spindle nose, and which has a disc E, at its outer end, slotted on the face to receive a crank pin, adjustable for radius, to give different lengths of stroke to the ram F, connection being made through the rod G. The ram slides between guides, one of which has a take-up strip.

The tool, of circular section, fits the hole *a*, a key in the tool shank fitting the groove seen in *a*, to fix it in its correct position. After being set, the lug *b*, slewing on the pin *c*, is brought round and over the tool to receive the thrust of the latter; *b* is locked by the spring plunger entering by its pointed end a countersink in the pin *c*.

A slotting attachment by R. M. Clough, of Tolland, Conn., is shown in Fig. 71. A is the main milling and drilling spindle of the machine, which carries a four-threaded worm B, driving the wheel C, which reciprocates the connecting rod D, and thence the ram E. The object in having a four-threaded worm is to reduce the speed of the ram for slotting. As the wheel has forty-four teeth, it is rotated but once to eleven revolutions of the spindle. The tool holder F is pivoted on a pin at *a*, which lessens the dragging of the tool on the return stroke, a spring above pulling it up to its bearing in the ram ready for the next cut.

In the machine the milling spindle A is carried by a sliding head moving on the face of the machine, similarly to many drilling-machine spindles. It is racked vertically by the handle G, operating an enclosed pinion and rack on the sliding sleeve. The bearing is split to take up the wear of the sliding sleeve. The handle H throws the worm wheel out of action when slotting is not being done.

The Overhanging Arm.—The type of overhanging arm in the pillar machines considered in this chapter, which is very generally adopted, follows in this respect the early Brown & Sharpe model. It is fitted into and pinched in a socket above the cones. The advantage is that it can be reversed end for end, turned aside, or taken right away. These are conveniences, but some rigidity is sacrificed thereby, and some trouble in readjustment for centres. It is, when properly fitted and pinched, a very satisfactory fitting, but only when so made. It is so seldom that it is necessary to remove the arm itself that a good many firms think it better to make it a rigid portion of the machine framing. In these types the overhanging arm, cast with the headstock, is a rectangular bar, grooved to receive an adjustable stay, with movable centre. The adjustment of the stay is strictly lineal, and it is therefore not liable to get out of line with the main spindle. It is clamped to the arm in any position required. In another type the headstock cap is prolonged to form a straight overhanging, turned arm, and a socket, sliding along this, carries the adjustable centre for the arbor. In this the socket can be turned out of the way, but the arm remains fixed. This method is adopted in the No. 4 B. & S. machine, the heaviest of their universals, and on some of the heavier plain machines. Still the arched arm is in greatest favour, because it is less in the way when swung aside than a rigid arm cast with the head, and it can be removed entirely. When a cross bracket is bolted up in addition, connecting the arm to the knee, the arbor is absolutely rigid under the heaviest cutting.

In the universal machines, not being designed for the production of the heaviest kind of work, but more specially for light milling, gear cutting, fluting, and grooving, the overhanging arm is often left unsupported. For the heavier classes of work, however, for which the plain milling machines are better adapted, support is commonly afforded to the outer end of the overhanging arm by means of two crossing slotted ribs, pivoted against the knee below, and bolted to bosses in the end of the arm above. These have a very slight appearance, but their triangular disposition affords ample stiffness under heavy cutting. Examples of other supports, more or less substantial, are given in succeeding pages.

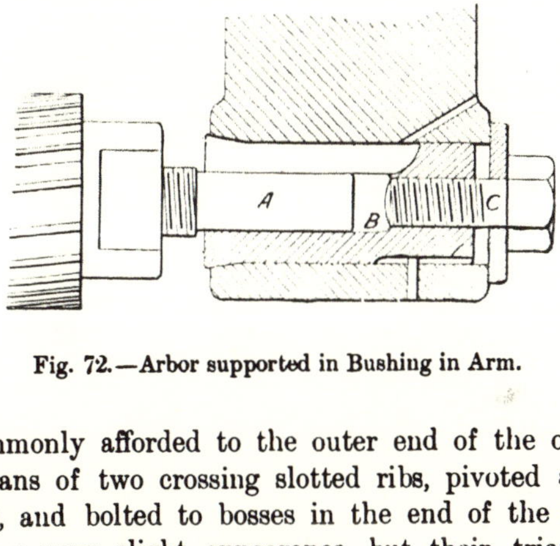

Fig. 72.—Arbor supported in Bushing in Arm.

One reason, perhaps the principal one, why overhanging arms are nearly always now made of circular in preference to rectangular section, is that the former is the better design for resisting all the stresses which come upon it from all directions.

Fig. 72 illustrates an alternative to point centres for affording support to the outer end of the arbor or mandrel. This is a parallel end A, encircled by a split bush of phosphor bronze B, fitting with a parallel body, and short tapered neck into the hole in the arm. It is drawn in by the screw C entering into its end. In this example three key-ways are milled in the bush, so that it can be turned one-third round to equalise wear.

Bracings.—The pillar and knee type of milling machine is seldom made without some form of bracing to tie the knee and the overhanging arm together. However rigidly the knee may

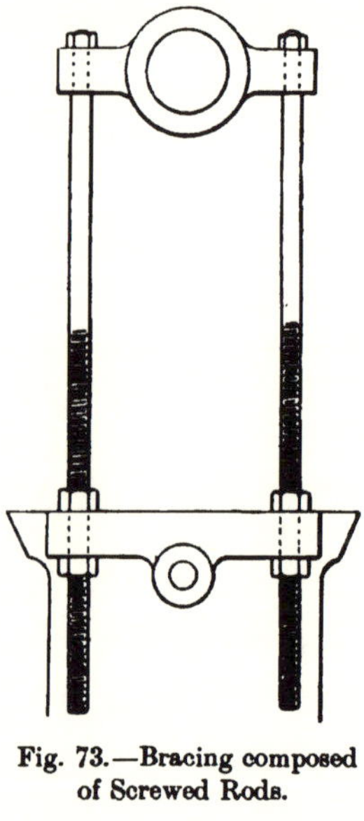

Fig. 73.—Bracing composed of Screwed Rods.

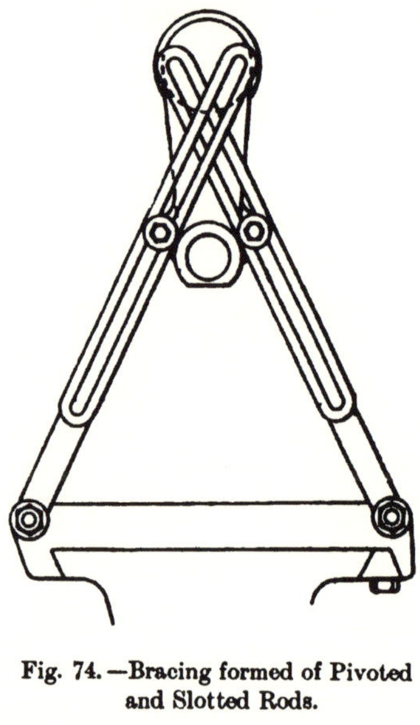

Fig. 74.—Bracing formed of Pivoted and Slotted Rods.

be built, it is nevertheless unsupported on one side, and there is under heavy cutting operations some tremor perceptible, which leaves its result in more or less chatter. For this reason some firms prefer to abandon the overhanging design and provide a solid base under the table, as in the Richards' design, making the cutter spindle adjustable vertically.

But the greater number of firms retain the convenient form of the open side table and vertically adjustable table and knee, using bracings in some form or another, which are more common on the plain machines than on the universals, the reason being that they have, as a rule, heavier duty to perform. For a large number of operations on the latter the bracing may be dispensed with.

A simple form of bracing is shown in Fig. 73. Two screwed rods are brought down from lugs in the outer arm support through similar lugs provided in the front of the knee. After adjustment is made for height, the double nuts will securely tie the knee to the overhanging arm above.

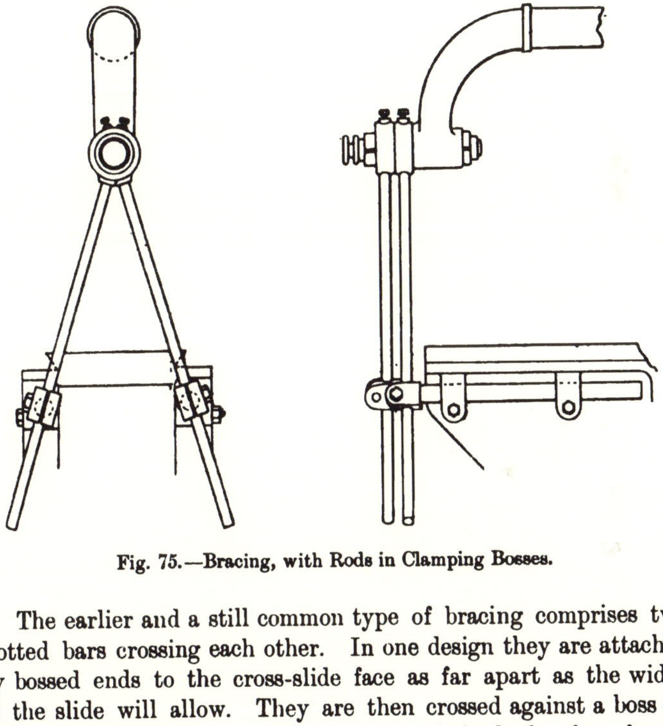

Fig. 75.—Bracing, with Rods in Clamping Bosses.

The earlier and a still common type of bracing comprises two slotted bars crossing each other. In one design they are attached by bossed ends to the cross-slide face as far apart as the width of the slide will allow. They are then crossed against a boss on the overhanging arm, to which they are bolted, the three bosses forming a triangle.

In another form, Fig. 74, two bosses are provided on the overhanging arm, so that each brace has its separate bolt, and the braces cross higher up. In each case adjustment for height is obtained through long slots in the braces, which slide over the bolts.

An objection to braces of this kind is that they encroach slightly on the cross traverse of the table, and in some cases they prevent the putting on of wide pieces of work overhanging the table, which it would often be desirable to do, but which the presence of such bracing prevents.

In several Continental machines the bracings provided are different from any of those made in England or America. Two round rods, Fig. 75, with bossed ends, are in some cases used, these fitting and pivoting over a large pin in the overhanging

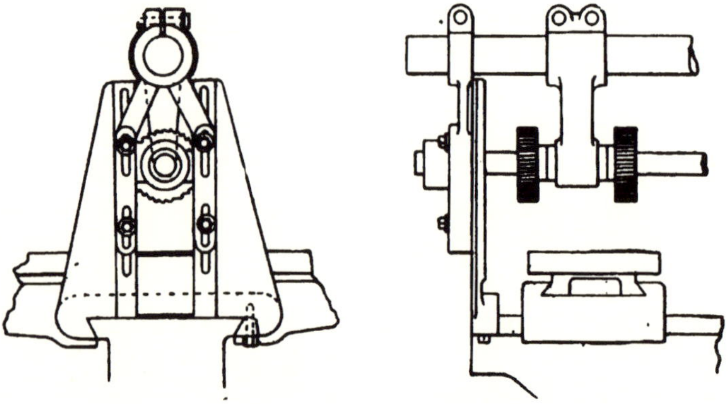

Fig. 76.—Double Cast Bracing with Slot-hole Attachments.

arm. The free ends of the rods go through split lugs at the ends of other rods, which lie horizontally along on each side of the knee, passing through split lugs on the sides of the same. Both sets of lugs have clamping bolts, so that, at whatever angle and height the braces are set, they are readily secured. A wide range of horizontal adjustment of the rods and braces is obtainable in this design to permit of the insertion of broad work.

Bracings of wrought iron or steel rods are not, however, so rigid in themselves as it is sometimes desirable that they should be. The methods by which they are attached make rigid connections, but the bars themselves are subject to some vibration when heavy cutting is being attempted. Theory indicates, and experience is in favour of cast-iron braces, the substance of which is more rigid and solid, and in which there is a better chance of massing and proportioning the metal.

One of the stiffest bracings of this type is illustrated in Fig. 76. The vee'd edges of the knee are embraced by a vertical standard, the two slotted uprights of which are connected at the bottom with a deep cross bar, but having a space above which

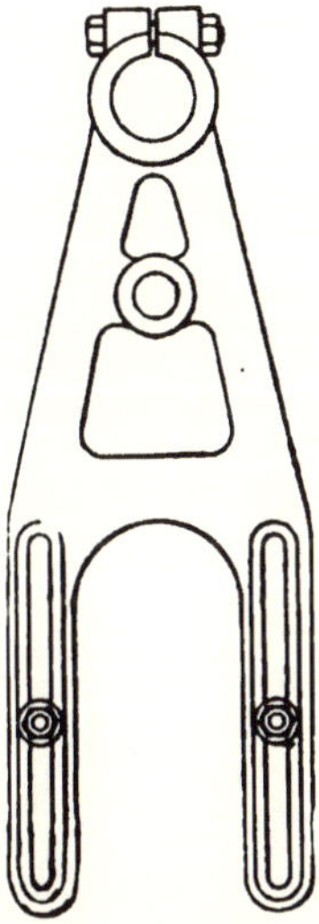

Fig. 77.—Bracing screwed to Knee with Bolts, with Adjustment Slots.

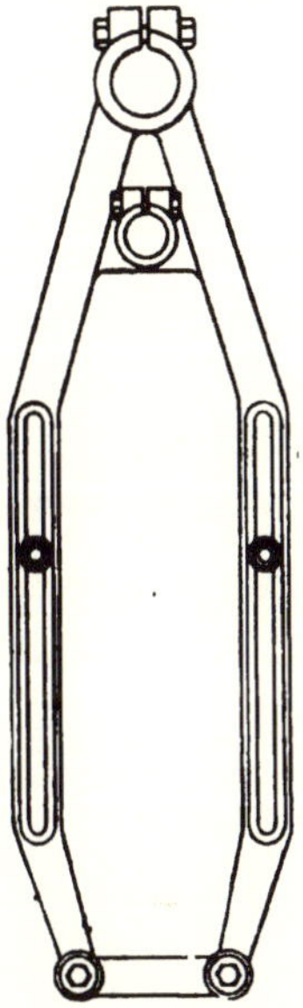

Fig. 78.—Deep Bracing, on Arm, Knee, and Base, with Slot Adjustments.

clears the overhanging arm when the knee is raised to its higher positions. The end of the overhanging arm is embraced by the boss of another bracket clamped on the arm, and having horns with slot holes coincident with those in the lower bracket. The two are bolted together at any vertical position of the knee, with four bolts. The horns of the upper bracket are tied together with a cross bar, and a boss serves as an outboard support for arbors. Such a bracing as this is perfectly rigid, making of the knee and is attained in many different designs, according to the ideas of different firms.

The Work of the Pillar and Knee Machine.—It would be more easy to say what this type of machine cannot do, than what it is able to do. Its only limitations are those of dimensions, for within its capacity, it is, in its highest developments, of universal range. For the type includes both the plain and universal designs, while in most of the later machines some provision or other is made not only for horizontal, but also for vertical and angular cutting (see page 75). Besides this, the type is made larger and stiffer than formerly, and better braced, so that, subject to its limitations in capacity, it is the most generally useful machine in the shop.

In this machine the height of the spindle being unalterable, the feeds are imparted to the knee, and all that it carries in the form of table slides, gears, &c. This taxes the elevating mechanism, and requires good fitting if unsteadiness is to be avoided.

The alignment of the outer support of the spindle bearing is embodied in the machine. The task of the workman therefore is confined to setting the work truly in all directions, and attending to the feeds. Methods of holding are illustrated in later chapters. Plain edge and profile milling are the special work of the machine, but a good deal of face milling is also done. The universal machine will do all that the plain miller will do, but the utilities of the latter are limited by comparison with the former. The plain machine being, however, built more heavily of the two, is preferably to be selected for heavy duty, and the universal for lighter, and for that specially requiring the swivel table and spiral head. Work requiring the dividing head alone, without the spiral movement, is better done on the plain machine if heavy cutting is desired. The functions of the two, therefore, notwithstanding their general resemblances, are kept distinct in shops where economical considerations are allowed to prevail.

On the universals there is no operation of the machine shop which cannot be performed, not merely as a makeshift, but as a perfectly fair and legitimate function. This, of course, excludes the work of the turning lathes. It is also true to say that any operation of the machine shop can be performed on the lathe if fitted with suitable appliances and adjuncts. But much of this is not legitimate work for the lathe; in fact, many jobs when so done are not done economically, but from necessity or through force of circumstances and conditions. On the milling machine it is all regular work, for which the machine has been specially designed. The exception noted to the work of turning is not absolute either, since wheels with very light rims are often milled more rapidly and more accurately than they could be turned in a lathe. The universal machine, therefore, fitted with suitable adjuncts, is capable of milling in rectangular relations, or at an angle, of milling circular outlines, or spiral flutes in a wide range of pitches, of cutting spur, bevel, worm, and rack gears, and cams, and all with a degree of precision for which little or no provision is embodied in other classes of machines, some of the most modern lathes and gear cutters alone being exceptions. The universal machine, as made by a few firms, is about the most beautifully designed and fitted article in the machine shop.

CHAPTER IV.

VERTICAL SPINDLE MACHINES.

Vertical *versus* Horizontal Spindle Machines—Vertical Spindle Machines—The Profiling Machines — Various Examples — Designs of Vertical Spindle Machines—Built after the Model of Drilling and of Slotting Machines—Examples—Spindles coming from Below—The Work of the Vertical Spindle Machine—The Work of the Profiling Machines.

Vertical *versus* Horizontal Spindle Machines.—A question that often arises is the relative utilities of vertical and horizontal spindle machines, and also the relative values of face and edge cutters.

The principal advantage of having the face of the work lying horizontally is that it is under observation, and that heavy masses can be handled and set better, as a rule, in that position. Fastening flatwise directly to a table is usually more convenient than bolting to the vertical faces of an angle plate. With a vertical spindle, both faces and edges can be milled with face and edge cutters respectively, without re-setting the work. This is not so conveniently done with horizontal spindles. Hence we find that, plano-millers excepted, the vertical spindle machines are more frequently employed than the horizontals for general engineers' work. The Lincoln miller is used more largely, in some of the lighter industries, but not so in engineers' shops. Besides which it was the first in the field, and has therefore had more time to become established.

Face and edge milling are often employed indifferently on vertical and horizontal faces. An advantage of working on vertical faces is that the chips fall away at once. On horizontal faces they should be swept off with a hand brush. Sometimes a suction pipe or a blast pipe with compressed air is used for the purpose.

Whether the spindle of a milling machine shall be horizontal or vertical depends mainly on the class of work for which it is selected. On the universal machine the horizontal position is preferable, because it is more convenient for gear cutting and fluting, and for work done by form cutters with linear traverse. The lubricant, too, lies on the work more efficiently under horizontal cutters. Generally, for broad-edge cutters also, the horizontal position of the spindle is best.

Vertical spindles are generally preferable when end mills are used. Profile milling must generally be done with vertical spindles.

On the other hand, the heaviest work is performed with horizontal spindles on machines of the planer and slabbing type. Better support can be afforded thus to the arbor, and therefore longer arbors can be used on the horizontal than on the vertical spindles. The latter are frequently supported at the lower end for heavy work, but not so efficiently as the former can be on their sliding heads on stiff uprights. Heavier cutting can be done on the planer type than on the vertical type.

Vertical Spindle Machines.—The pillar and knee type of machine has to be greatly modified when a vertical spindle is employed, and dimensions increase. The knee, always a weak element, is then frequently abandoned, the base of the pillar being extended—in slotting-machine fashion—to carry compound tables having no vertical feed. Then the vertical feed is imparted to the spindle and spindle head, and these are counterbalanced. The machine thus much resembles in outline a powerful drilling machine. Frequently also geared drives are abandoned in favour of the belt drive over guide pulleys.

These machines include examples of the heavier and the lighter types, constituting a large group, the members of which are employed more extensively than others, the Lincoln, and pillar and knee excepted. They have both their advantages and disadvantages. They take the place of the slotter in tooling circular pieces, and perform the work more correctly, because the slotting tools produce minute facets, leaving finish to be imparted in other ways. The vertical miller tools internal and external portions on a piece without re-setting, or on segmental pieces arranged in circular order. Both faces and edges can be tooled without re-setting—faces with face mills, and edges with edge mills.

It is also the best type of machine on which to fit profiling attachments.

The disadvantage of the type is that due to milling on horizontal faces, which does not permit the cuttings to fall away as they do from vertical faces tooled from horizontal spindles. And when vertical faces are being tooled on it with edge mills, limitations of depth come in, due to the spring of the arbor. This is controlled in the heavier machines by a bottom supporting bracket, which fulfils the same function as the outer arbor support on the horizontal spindle machines. This support is variously fitted, being sometimes brought down from the lower spindle bearing, sometimes hinged at the side of the upright. But the essential is that it shall have provision for being swung aside, to permit of doing work for which the support is not required.

Within the very comprehensive generalisation of the vertical machines there is included a large number of designs, varying in the methods of the drives, through gears or belts, with vertical adjustments imparted to tables or to spindles, and with heads having one range of movement only, or combining horizontal or angular settings.

The Profiling Machines may be common vertical spindle machines first, to which profiling movements are added, or the profiling may be the principal function, plain milling being a secondary thing. The first division generally embraces the larger machines, the second the smaller ones.

In profiling machines the table is not adjustable vertically, but that movement is effected in the spindle. The machines in some cases have rigid knees, on which the main table slides, in others a bed of greater or less length carries the table, which then resembles the general type of planing machine, and the spindle slide is carried on a cross rail adjustable on the faces of housings attached to the bed.

Methods of profiling are described later. The essential of a profiling cutter slide is that it shall be coerced, not by screws, but by the pull exercised by a weight suspended from a chain. The friction of the slide is lessened as much as possible by fitting rollers to the lower or the upper bearing edges, and steadiness is secured by giving a long contact between the sliding parts. The former pin may or may not have a range of adjustment. The spindle is driven directly by belt in the smaller machines, through cones and back gears in the heavier. The mass of the spindle and its bearings is counterbalanced. Though most profiling machines have vertical spindles, an exception occurs in some made by Messrs Webster & Bennett, of Coventry, which have horizontal spindle heads resembling those of lathes, on a lathe form of bed.

Small machines for general milling and profiling have vertical spindles of from $\frac{3}{4}$ inch to $1\frac{1}{4}$ inches diameter. A horizontal cross-head is cast with two uprights, and the latter are bolted to a box base which carries guides for a table, upon which the work is traversed under the mill. A carriage slides transversely along the crosshead, adjustable by handle lever, and the spindle slide moves vertically on the carriage, its movement being adjustable by means of vertical stops. In these the spindle is revolved by means of a belt and drum only. The traverse of the carriage, and also that of the table, is effected by hand. Light double-spindle profiling machines are also made nearly similar, so far as the methods of operation of the slides by hand are concerned. The traverse movement of the slide is effected through spur gear and a rack on the bottom edge of the slide. The driving of each spindle is from a long drum at the rear, driving half-crossed belts on to the spindle pulleys, the drum being fitted with fast and loose pulleys. The spindles are balanced with springs above.

A vertical spindle machine of simple design, by Messrs Webster & Bennett, is shown in Fig. 82. The style of frame provides for carrying the table slides without vertical adjustment, this being imparted to the spindle slide only.

The method of driving the spindle A is by the belt pulley B, which runs upon a sleeve, in order to relieve the side pull; to B the belt is carried over idler pulleys C, from the driving one D. A four-stepped cone on the same spindle as D provides for variations in speed. Some types of this machine are without gear, but others have a back gear contained within the pulley B.

The spindle is moved up or down by the slide E, which is travelled by the balanced handle F. On the same shaft as the latter is a worm, which drives a wheel working upon a vertical screw, so imparting a fine feed to the slide E. As the lower end of the spindle A is confined in E with collars, it partakes of the movements up or down, the upper splined end passing through the pulley B, in which keys drive on to the spindle, while still permitting the latter to slide endwise. A stop screw G, in a boss on E, abuts against a lug standing out from the fixed frame, seen

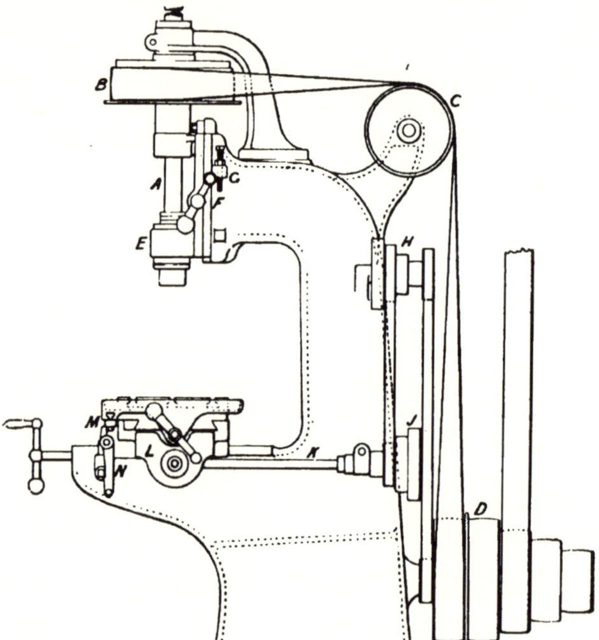

Fig. 82.—Vertical Spindle Machine.

in the figure, and so provides for a definite depth of feed, which may be repeated as often as desired on repetition pieces.

The table movements are effected by the handles seen, or the longitudinal feed by power through the four-stepped cone H, driven from a pulley forming an extension of D. H imparts motion to another cone J, revolving a shaft K. The latter drives a worm beneath the apron L, engaging with a wheel which is splined on, and rotates the feed screw, on the end of which the balanced handle is seen.

An automatic knock-out is provided, which drops the worm

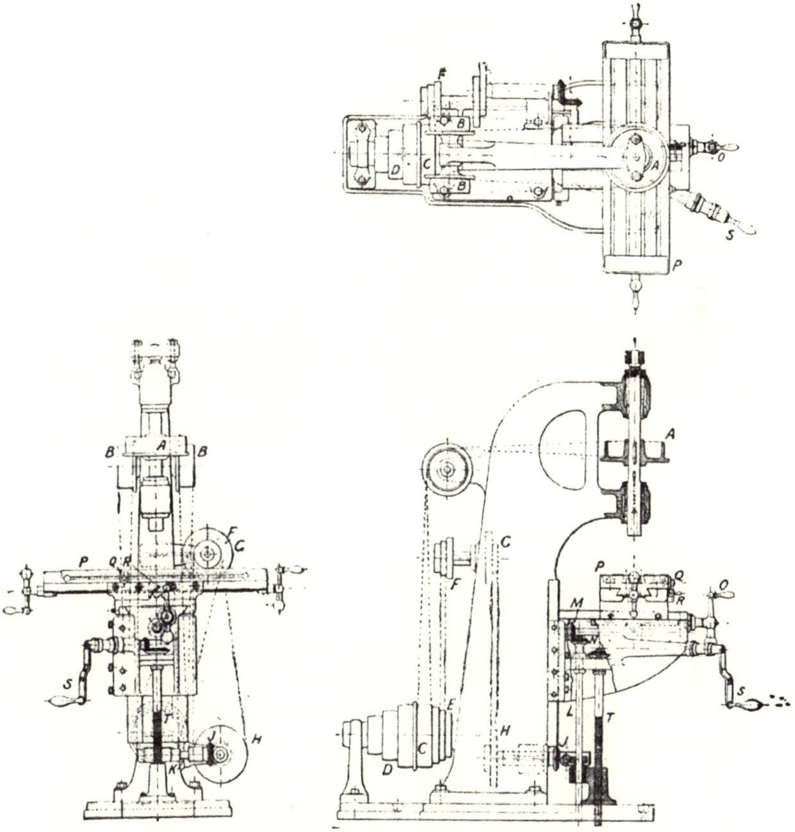

Fig. 83.—Vertical Spindle Machine.

out of engagement with its wheel at a predetermined position of the table travel. This is effected by dogs bolted on the under ledge of the table at M. When one of the dogs strikes against the end of the lever N, the latter, which normally holds the worm in a hinged box up to engagement, allows it to drop, so letting the teeth out of mesh, and stopping the feed. The cases where this feed are used comprise those in which a long cut is taken, and the stop is set to trip the feed at its termination. For short work the hand feed is employed.

Fig. 83 is a vertical spindle machine by J. Parkinson & Son, which embodies vertical movement to the table instead of to the spindle. The latter is driven by a plain pulley A, through idlers B, B off a pulley C, the latter being given four speeds through the cone D on the same shaft.

The tables are operated through the handles seen, or the longitudinal and cross movements are actuated by power. The feed is derived first from the three-stepped cone E, on the same shaft as C and D. E drives up to another cone F. On the same shaft is a two-speed pulley G, driving down to a similar one H. It will be seen that, by means of the two sets of cones, five rates of feed can be obtained for each rate at which the spindle may be driven.

From the shaft of H a pair of mitre gears J drives a short horizontal shaft K, upon which a worm and worm wheel convey the motion to a vertical shaft L, mitre gears M at the top of which drive a horizontal shaft within the knee. The shaft L is splined to slide through the worm wheel, to accommodate itself to the vertical table movements. The shaft N, driven by the mitres M, first rotates the feeding screw above it (on which is the handle O) through a pair of gears, and also the screw in the longitudinal table P through mitres, seen dotted in the views. A trip device is fitted to the table P, seen in the front elevation, the block Q striking a lever R, which disengages the feed within the table. The vertical travel of the knee, and table is effected by the handle S, working bevel gears which rotate the vertical feed screw T, turning in the threaded foot at the base.

A profiling attachment is made for use on this machine, which embodies the principle of the former or templet, controlling the movements of a roller. Three views of this device are seen in Fig. 84, in plan, and front, and side elevations. A circular revolving table A is driven by worm gear through spur wheels, actuated by another worm gear, which derives motion from a telescopic shaft, the latter being employed because of the irregular movements of the table under the guidance of the former. The work is carried upon a small table B, the edge of which is shaped to the required outline to be milled; a roller C presses against the edge, being held in a bracket D. The holder of the roller C is capable of adjustment with a screw E, to permit of milling sizes larger or smaller, and giving the depth of cut. The sliding table A, and small table B, are pulled against the roller C by weights F.

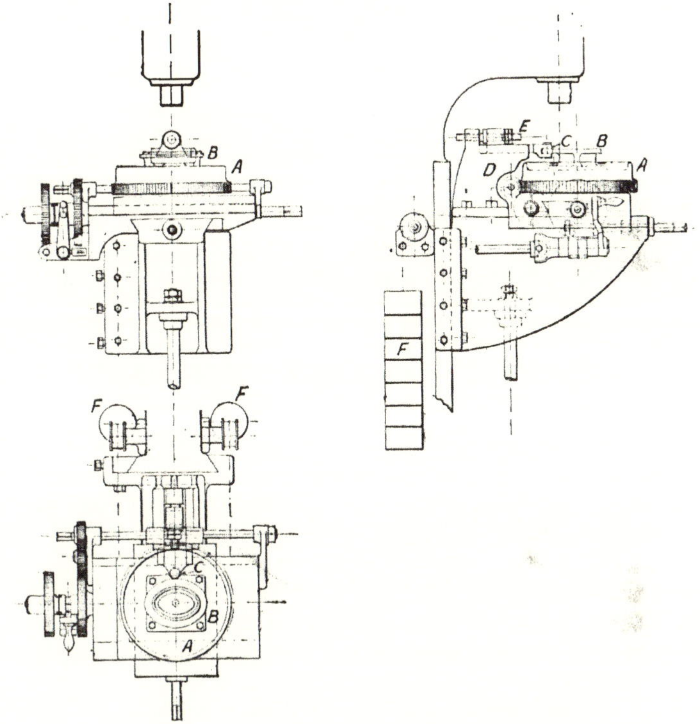

Fig. 84.—Profiling Device for Parkinson's Machine.

It will be seen therefore that on revolving the table A, it will partake of the motion given by the profile of B (which is that of an ellipse in the illustrations), and so bring the work likewise against the cutter.

A circular table, also fitted to the machine, is shown in Fig. 85. It is bolted down to the tee slots of the main table, and is provided with a worm and hand wheel, which rotate it. Graduations are made around the edge. The worm can be thrown out of engagement, to enable the table to be rapidly turned for purposes of adjustment. It can also be locked when straight portions have to be milled, the circular table being then carried along bodily by the machine tables.

A vertical profile milling machine by Messrs Webster & Bennett of Coventry is shown in Fig. 86.

The drive to the spindle is by belt from a drum at the rear, which is necessarily made long to permit the belt to follow the cross traverse of the carriage C along the cross rail D, in the act of profiling. The carriage runs on rollers on the top of the cross rail, in order to reduce the friction as much as possible. The spindle slide E has a vertical adjustment on C. The former pin is seen at *a*. This is held closely against the form or pattern by the pull of the weight seen at the side, which draws the slide C by the chains indicated, passing over pulleys. The saddle C can be clamped on the cross slide D when ordinary milling has to be done.

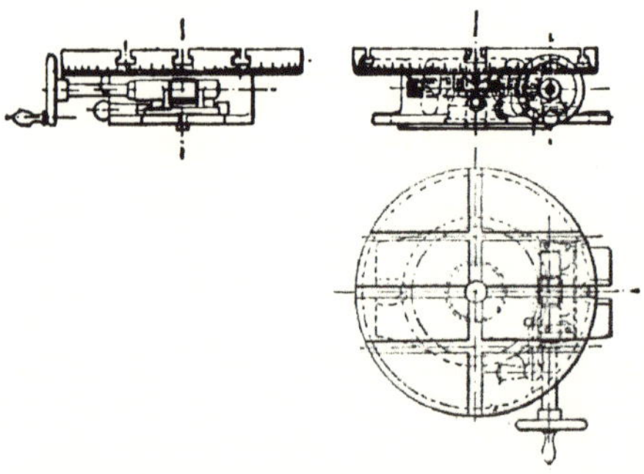

Fig. 85.—Circular Table for Parkinson's Machine.

The table F has a longitudinal feed, that is, perpendicularly to the faces of the slides, derived from the stepped cone G, through the train of spurs, and worm gear shown, to the rack beneath the table. The stops *b b*, on striking the lever *c*, reverse the motion through the claw clutches *d*, on the worm shaft H. A handle put on H affords a hand feed to the table.

The table also has a circular feed through worm and spur gears, which can be disconnected instantly by throwing out the clutch *e*. A pump J is belted from the main driving shaft. The body of the machine forms a sud tank.

With increase in the diameter of spindles, the profiling machines approach more nearly in the length of table and bed, and in general appearance to the planing machine framing. A circular

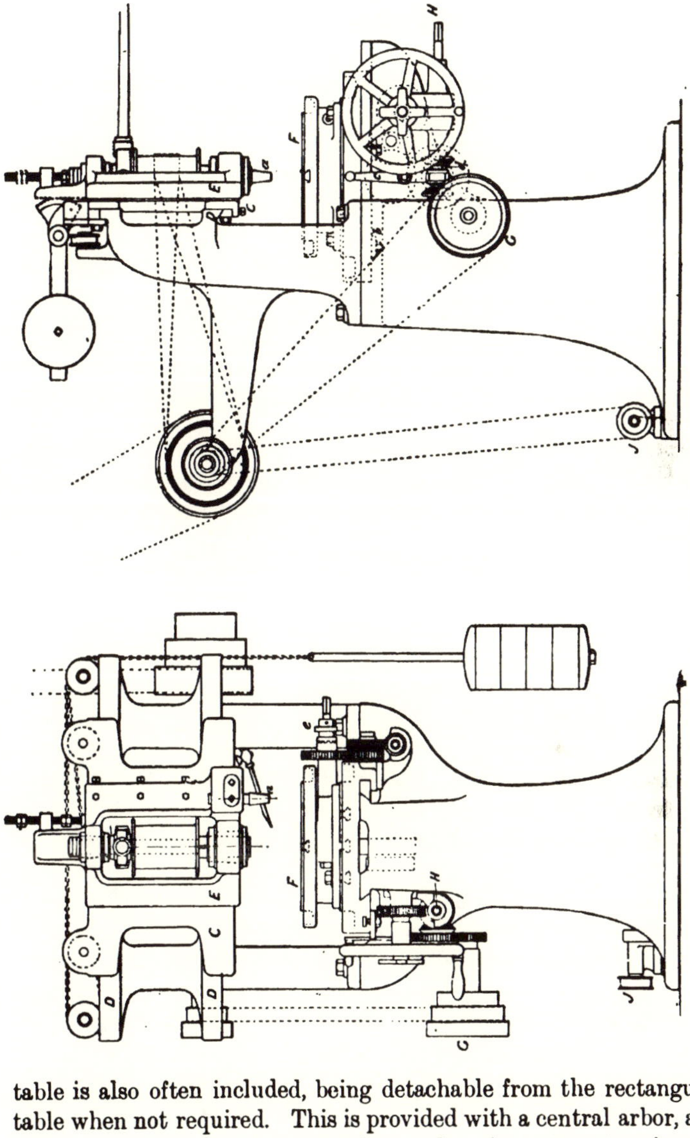

Fig. 86.—Vertical Spindle Profiling Machine.

table is also often included, being detachable from the rectangular table when not required. This is provided with a central arbor, and with worm gear for rotating the work. A cutter stay is also attached, being made so that it can support the lower end of the mandrel or arbor in any position. It is attached to and swivelled upon the spindle slide, and locked in any position. In some machines the slides are indexed with steel rules, and the circular tables are also indexed. Machines of this kind are made with spindles up to 6 inches diameter.

As the capacities of these machines increase, nearly every detail becomes much modified. The hand lever, by which the cross traverse of the spindle slide is adjusted in the light machines, gives place to a weight, the chain of which passes over pulleys at the head of the framing. Tables are extended, so that, instead of getting a table traverse of from 12 to 18 inches, we have traverses equal to those of the average planing machines, so that objects can be profiled up to 4 ft., 5ft., and 15 ft. in length. The tables are, in fact, like those of planing machines, made with numerous tee slots, and running on flat beds with vee'd edges. They are driven by screw, have self-acting feeds, reversals, and quick traverses by power. Troughs around the edges are provided to catch the oil. The feeds of the traverse carriage, though still capable of being operated by hand, are also rendered self-acting, and variable, and reversible. Again, though machines of very considerable dimensions, up to those with 2-inch spindles, are driven by belt, the spindles of those of large sizes are driven through stepped cones and bevel gears. In some machines, again, the cross slide is not confined horizontally by the uprights, but slides vertically upon them, exactly as in a planing machine, being adjustable by hand or by power. To all machines of this kind a centrifugal pump is fitted for the lubrication of the cutters.

These types of machines are equally suitable for the work of general milling, the profiling attachment being a convenient adjunct superadded, and one which extends the usefulness of the machine, without detracting from its value as a plano-miller.

The principal vertical spindle machines may be broadly classed as those having tables fixed in regard to height, and those in which the tables have vertical adjustments and feeds.

The first-named great group have framings much resembling those of the slotting machine, or the similar outline of drilling machine framings. There is a broad base, usually cast in one piece with the uprights, with slides to carry the table slides, and an overhanging head, either cast with the upright, or bolted to it. The spindle bearings are movable vertically, and the lower bearing, or the entire bearing arrangements have vertical adjustments, not only to suit the height of the work, but also to afford support to the cutter close to the latter. The spindle ranges from 3 in. to 5 in. diameter. The spindle slides are counterbalanced in various fashions, and are adjustable by hand, and also by power in the heavier machines, the latter being a slow motion by a belt-operated worm and a wheel. The spindle drive takes place through stepped cones, with back gears included, and through bevel gears for changing the direction of motion in all the older, heavier machines, and in the majority of the recent ones. But a considerable number of the latter are now fitted with high-speed belt drives, even in heavy machines. The belts then come over guide pulleys, driving immediately to the spindle, which can be operated directly, or through back gears. Bevel wheels are eliminated in this design. The pull of the belt does not take place directly on the spindle, which would produce unequal wear in the bearings, but on a sleeve which encloses the spindle (see page 109).

Feeds are taken from belt cones, and all the table motions are self-acting, and controllable by the attendant at the side or front of the machine. A circular table is almost invariably included, also made self-acting through worm gears. The tables are compound; if the additional table is not circular, but square, it has the circular motion. The feeds, which are towards or away from the vertical frame, transversely thereto, and circular, are each self-acting and reversible through gears and screws, and worm and wheel, similar to those employed in slotting machines. The reversals are effected by hand, either through three bevel gears and a sliding clutch in many machines, or by means of a friction disc and roller. Each feed motion also is capable of disconnection by means of friction clutches, and of quick hand traverse. The circular table is in the way of fixing work for plain milling, and is therefore removable. But in some designs this is rendered unnecessary by the simple device of flanking the circular table with wing tables, which partake of the rectilinear movement of the slides below. The faces of these stand a trifle higher than the face of the circular table, so that work bolted to them clears the circular one, and can be thus tooled without removing the latter. On vertical spindle machines, fitted with compound slides and circular table, all rectilineal work can be done with face or edge mills; and all curves, alone or in combination, with straight edges.

It is an exceedingly useful type, better suited, perhaps, than any other to the requirements of the general shop. It covers a large volume of work, and will take heavier and larger pieces than the pillar machines previously noted. It is made in a wide range of dimensions. Some of the machines of this class which have been constructed in recent years are exceptionally massive and stiff, capable of doing plenty of heavy cutting. The details are worked out in many ways by different makers, so that beyond the main design there is not much in common between them.

An unsatisfactory feature of the ordinary slotting-frame type of milling machine is the projecting unsupported spindle, resembling in this respect the ordinary drilling and slotting machines. It is not possible in some jobs to use the bottom supporting bracket provided in many machines of this class. There are two methods in use, therefore, for supporting the spindle and increasing its stiffness as far as possible, one of which in its essential features is embodied in most American drilling machines, and on a very limited number of slotters.

In a vertical milling machine by Hulse & Co. Ltd., steadiness of the spindle against the pressure due to the stress of the cutting action is ensured by enclosing it in a long, hollow square slide, making it run in conical bearings therein. The slide is made capable of vertical movement, and can be set and clamped by a locking screw at any height required. Vertical movement of the spindle is provided for by means of a pinion that embraces it, and which is driven by a pinion long enough to cover the whole range of travel of the spindle. The weight of the spindle and slide is slightly overbalanced by means of a balance weight. Several firms, both in England and America, construct vertical milling machines of the slotter type in which the spindle head can be adjusted vertically by a hand wheel and gear to suit work of different depths.

In the second great group of machines the knee is adjustable, the arrangements then resembling those of the pillar and knee machines. The objections to this design increase with mass, and therefore it is not adopted much for the heaviest types, though some good massive examples occur. The vertical adjustments of the knee are not always employed for feeding as in the ordinary pillar and knee machines. Frequently it is a movement of adjustment only, the cutting feeds being imparted to the spindle, as being more convenient. These generally have a micrometric device, and in good machines an adjustable dead stop which regulates the depth of cutting for any number of similar pieces.

The three views, Figs. 87-89, give general and side elevations of a vertical milling machine by Messrs Alfred Herbert Ltd., of Coventry, which combines the best points of present-day practice. It is a massive tool, weighing about 6,000 lbs. The leading dimensions are:—Longitudinal feed of table, 36 inches; transverse feed, 12 inches; vertical adjustment of table, 12 inches. The spindle is $2\frac{5}{8}$ inches diameter. The circular table is 16 inches diameter. The maximum distance from the surface of the table to the spindle is 19 inches. Feeds are automatic, are reversible, and have automatic trips in both directions, and dead stops to table feed. Gears are all enclosed, and lubrication is amply provided for. The machine is of the type in which the knee is adjustable vertically, and the spindle has similar adjustment, the fine adjustments being given with the latter. The spindle is belt-driven, direct or through spur gears, no bevels being used anywhere in the driving mechanism. These are the leading elements, whence we now proceed to work through the details, enlarged views of which are given in subsequent Figs. 90-97.

In Fig. 88 the countershaft is driven by two sets of pulleys, A and B, at 375 and 150 revolutions per minute respectively. The single-shipper lever shown actuates each set of striking gear by being pulled into contact with either one of the two sets of blocks *a, a* on the shipper bars. The three-stepped cone pulley C drives to the spindle, the two-stepped cone D to the feed cones, and the pulley E to the pump.

Commencing with the spindle details, the cone pulley F is driven by C. As these have three steps, and there are two counter speeds, this gives six spindle speeds for either belt or back gear, so making twelve in all. They range from 20 to 500 revolutions

Fig. 87.—Vertical Spindle Machine. (A. Herbert Limited.)

per minute, the quicker speeds being reserved for fine finishing with the smaller cutters. The drive takes place from the pulley G on the cone shaft below, over the two guide pulleys H to the spindle pulley J. The latter, it is to be observed, is not keyed directly on the spindle, but its pull is taken by an intermediate bush or sleeve K encircling the spindle, relieving the spindle of

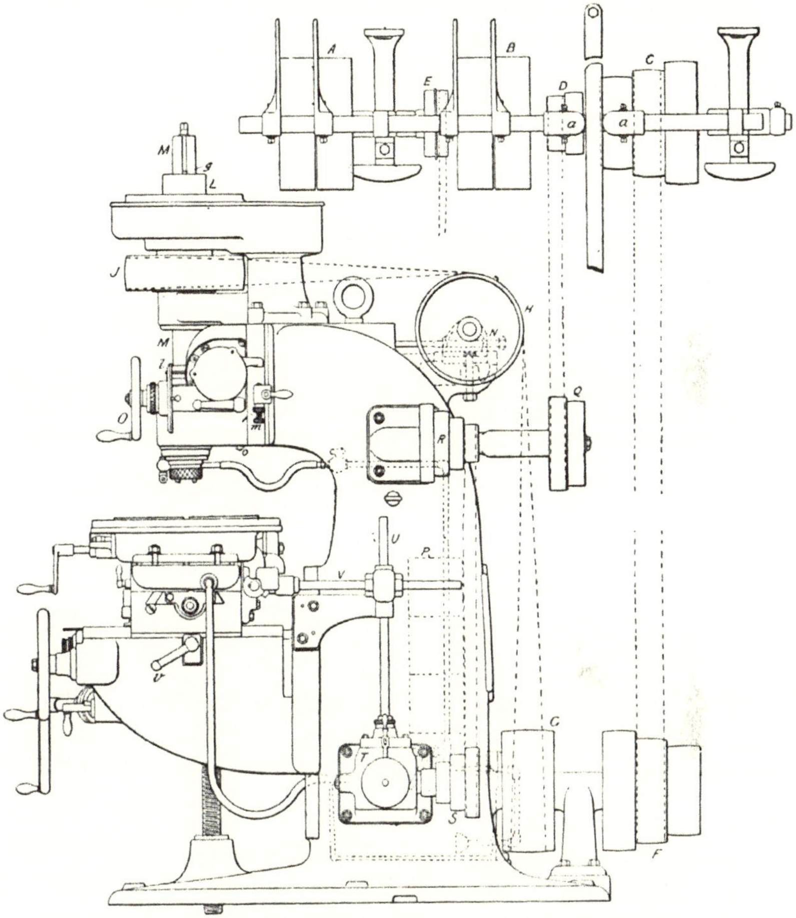

Fig. 88.—Side Elevation of Herbert Vertical Spindle Machine.

side pull, which would cause unequal wear. The driving of the spindle is transmitted as follows (compare with the details which are shown enlarged in Fig. 89):—

The boss of the pulley J is keyed on a boss encircling the sleeve

K, and this boss forms an extension of the first gear *b*. *b* drives *c*, with which gear *d* is solid, and *d* drives *e*. *e* is secured to a driver plate *f*. L is an equalising driver, whence motion is transmitted to the spindle by two keys *g*, *g*, to receive which the spindle is splined on opposite sides down a good portion of its length to allow of its sliding motion taking place. In this way the spindle

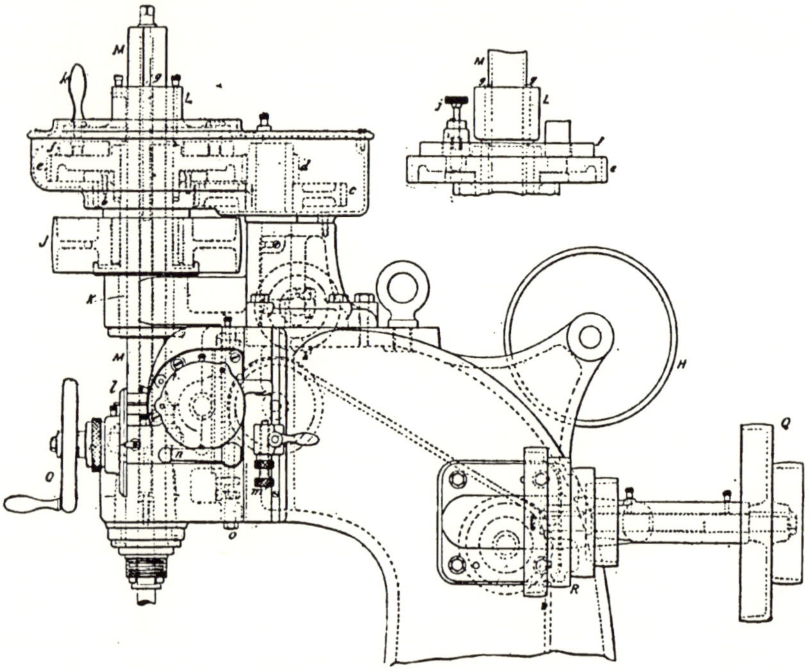

Fig. 89.—Details of Spindle Drive.

drive is given through the back gears *b*, *c*, *d*, and *e*. All these are enclosed by the light cast-iron casing shown in Figs. 87-89.

The back gears *c* and *d* are mounted on an eccentric spindle, which is thrown in or out by worm gear, indicated in the figure, and actuated by the hand wheel *h* to the left of the machine. Direct driving then takes place on locking the wheels *e* and *b* with a spring pin *j*, seen in the detail to the right of Fig. 89, the spindle being rotated by the driver L and keys *g*, *g* as before. The handle *k* in the figure is used to revolve the driver until the pin *j* comes over its hole, the casing preventing this operation being done by pulling on the wheel itself. The tension of the belt on the pulley J is maintained in an alternative design of the machine by means of the sliding bracket N, Fig. 88, and of its set screw in opposition to the abutment piece on the main framing.

A slide which carries the lower bearing of the spindle is capable of vertical adjustment to move the latter and afford continual support close to the cutter. Exact feeding is attained with the hand wheel O at the front, operating through a worm and wheel enclosed in the casing seen. On the same spindle as the

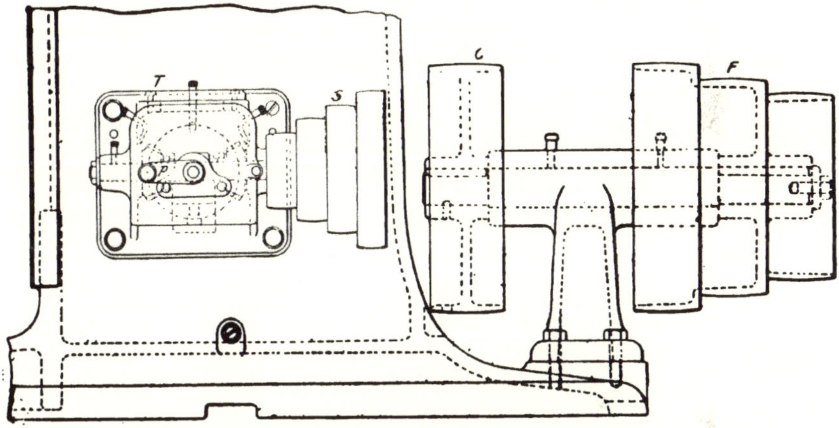

Fig. 90.—Side Elevation of Reversing Gear Box.

worm wheel is a pinion, which gears into a rack on the machine frame, and so racks the bearing up or down. A dial *l*, mounted on the hand-wheel spindle, is graduated into thousandths, and is adjustable to zero in any position, in which it may be then locked. The motion of the spindle bearing can be arrested at any height by the setting of the vertical stop *m*, which has fine screw adjustment and a lock-nut. A handle *n* locks the slide when face cutting is being done, and both this handle and the stop handle "lock" when in the horizontal position. The slide is balanced by weights P within the framing, the chain for which passes over guide pulleys, the anchorage being at *o*.

The conical spindle bearing is of hard phosphor bronze, with provision for taking up wear. The end thrust is taken by washers of hard tool steel and phosphor bronze. The spindle is of crucible steel, bored throughout. Its nose has a No. 10 taper, with a clutch drive for the arbors. The latter are retained in place with a draw bolt. The nose is threaded to receive large cutters, and when these are not in use the thread is protected by a cap.

Tracing next the table feeds from the countershaft pulleys D to the pulleys Q on the machine, these drive four-stepped cones R and S, giving eight feeds, doubled by the pulleys A and B on the counter. They range from ·500 to 12·75 inches per minute. The

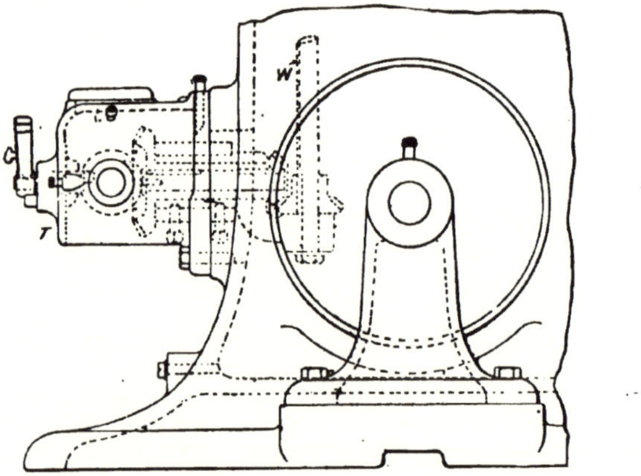

Fig. 91.—End Elevation of Reversing Gear Box.

feed motion for the circular table is transmitted through the reversing gear box T to the vertical shaft U, thence through spiral gears to the horizontal shaft V. A comparison of the views of this box in Figs. 90 and 91 will render much description unnecessary. The reversal is effected by the handle *p*, which slides a splined clutch into engagement with either one of a pair of loose bevels, so driving the wheel in the nest of three gears in one direction or the other in a manner well understood. The extreme positions of the handle are fixed by the studs outside the box.

The table feeds, derived from the shafts U and V, may now be studied in Fig. 92, compared with the general views, Figs. 87 and

88. The two views of the machines are not wholly alike, because the general views include a circular table which is not illustrated in the details; but that does not affect the main table drive, which we are now to consider.

Taking up the connection at the vertical shaft U, and horizontal shaft V, the function of these is to drive the circular table in the manner indicated in the side view, Fig. 88, but not shown in detail there. The drive of the oblong table takes place from a vertical shaft W within the framing, which is actuated from bevel gears direct from the gears in the reversing box (compare Figs. 91 and

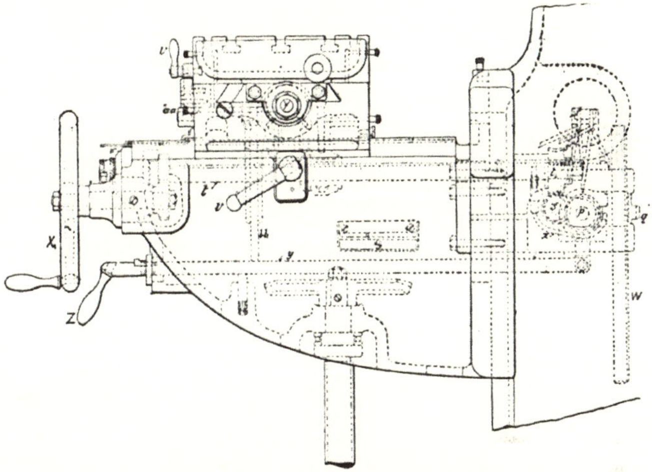

Fig. 92.—Detail of Knee, and Table Drives.

92 with Figs. 88 and 93). At the upper part of W a sliding spiral gear *q* is seen, which is the first element in the table feed. *q*, with a nest of gears is carried in a bracket which is bolted to the back of the knee, and therefore moves vertically with it, the corresponding sliding of the gear *q* being provided for in the splining of the vertical shaft W. *q* drives a spiral of equal size at right angles on the end of a shaft which carries four spur gears driving four smaller gears on a shaft *s*. On the latter, and mounted between the spurs, are two worms driving worm gears *s′ s′* on the shaft and screw *t*, *t*, the worms being double-threaded. The shaft

and screw *t, t* come out through the knee nearly to the front, Figs. 92 and 94, where, by means of a train of three spur gears each, the middle one being an idler, they are connected to the hand wheels X, X at the front of the knee. The hand wheels are thus separated sufficiently far to permit of both being retained on their shafts at one time, while the idler wheels allow the rotation of the hand wheels to take place in the same direction as the shaft and screw. As the worms can be dropped out of mesh with the worm gears, the shaft and screw can be worked either by power from the one end, or by hand from the other.

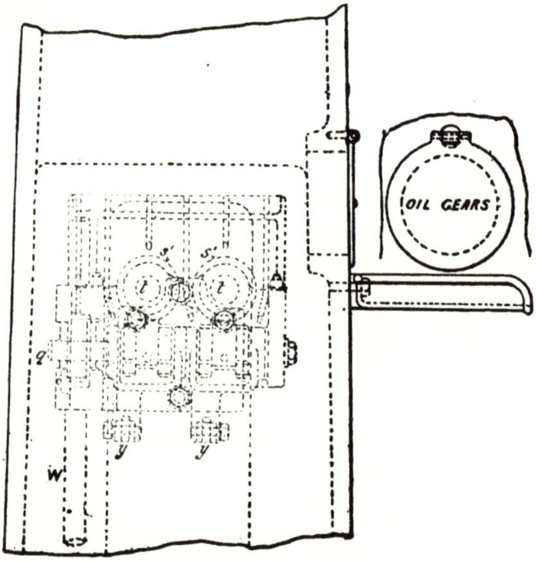

Fig. 93.—Detail of Driving Gears in Column.

The connection between one of the shafts *t* and the screw Y that imparts the longitudinal feed to the table is easily seen, being traceable through the mitre gears in Figs. 92 and 95. The transverse feed to the knee saddle is through the medium of the nut *u* seen in Fig. 95. The vertical adjustment to the knee is clearly seen in Figs. 92 and 94. The three screws for these motions are of the Acme form. The weight of the knee is taken on a ball race, and the crown bevel wheel is of large diameter. From these

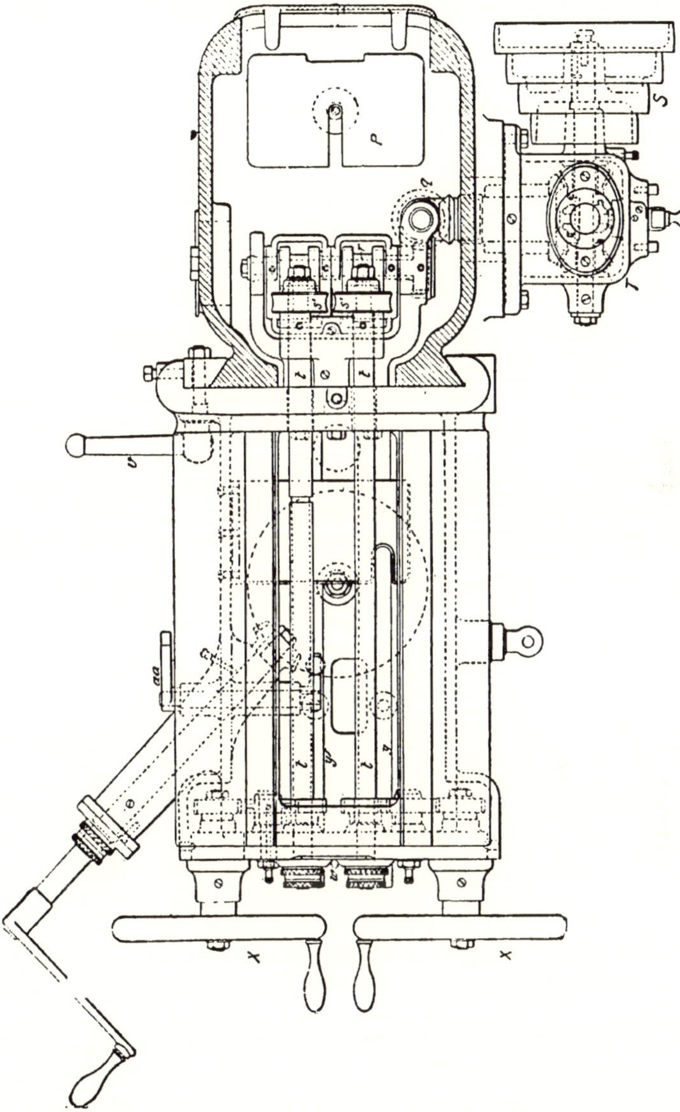

Fig. 94.—Plan of Knee, with Table Drives.

three movements we now pass to notice the locking and other arrangements.

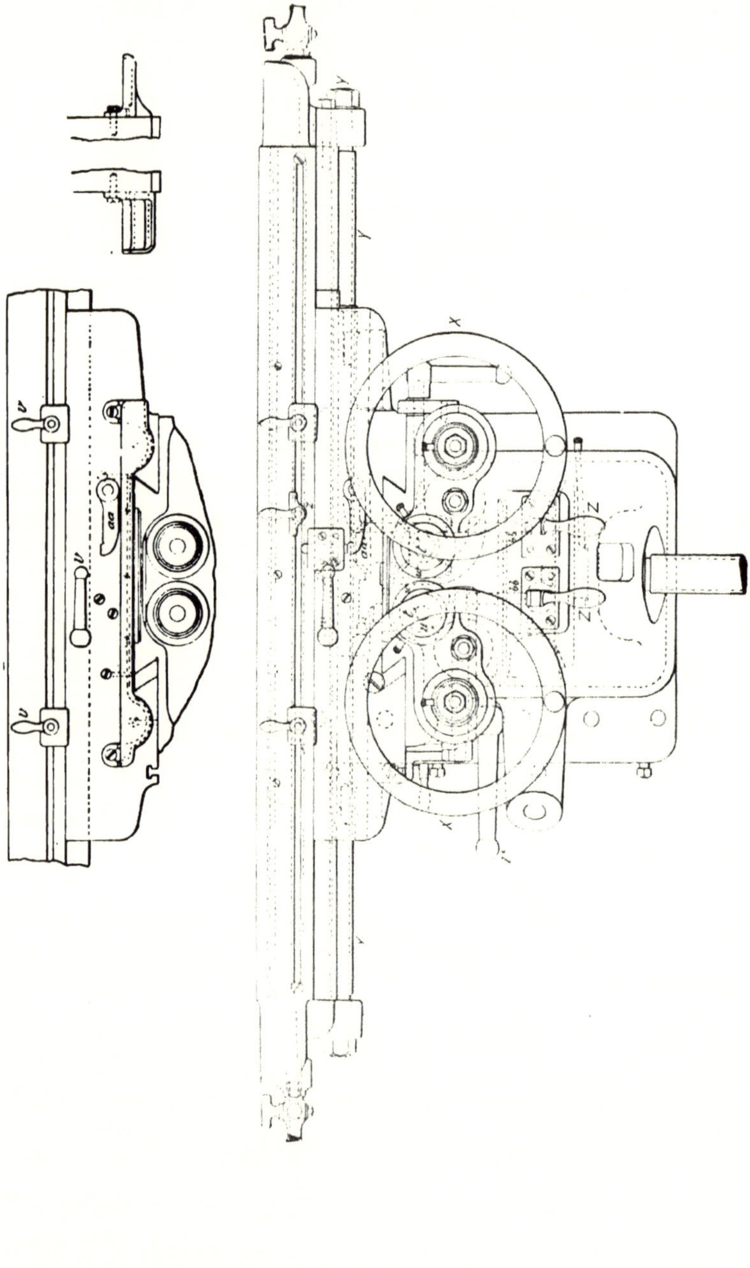

Fig. 95.—Front Elevation of Knee and Table.

It will be observed that every slide has its locking handle or handles. These include the table, knee, and saddle. The various locking handles are all distinguished by the letter *v* in the views (see also Fig. 96). All the feed screws have micrometer adjustments, the dials for these being indicated by the letter *w*, and one is shown enlarged in Fig. 97. The handles z, z, Figs. 92 and 95, are for dropping the worms on the shafts into and out of engagement with their wheels. These handles drop the box *x* which carries the worm shaft *s*. The box pivots around the axis of the shaft *p*, and thus the movement of z and the rod *y* throws it down or up, breaking or effecting gear between the worms and their wheels *s'*, *s'*. This is a spring plunger to cause the box to drop with certainty (see Fig. 92). The movement of the rod *y*

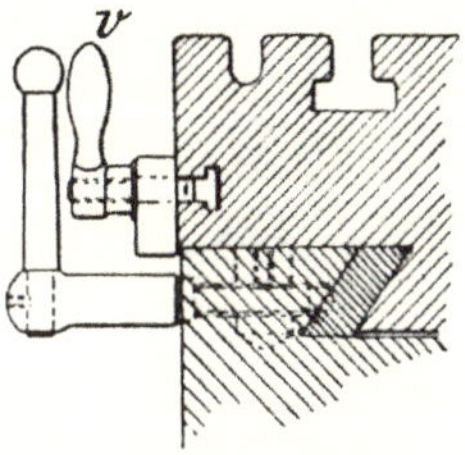

Fig. 96.—Locking Handle of Gib.

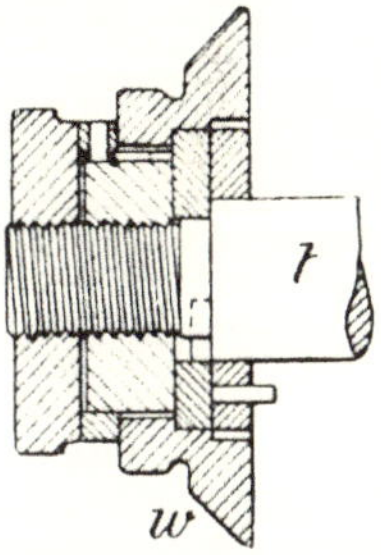

Fig. 97.—Micrometer Dial.

automatically, constituting an automatic trip, as shown in Figs. 92 and 95. The adjustable stop *z* clamped to the tee slot on the edge of the table, coming into contact with the top of the spring plunger, seen adjacent to it, thrusts down the lever *aa*, and with it a lever within the knee which bears on the upper ends of rods *bb*, Fig. 92. These press on the rods *y*, and so throw out the gear box *x* at any predetermined point. The spring plungers in the knee under the rods *y* serve to just hold up the latter from dropping when in gear and locked. In addition to these automatic trips, dead stops are fitted—blocks clamped in the tee slots of the table edges, and abutting solidly against the automatic plunger casing instead of passing over the rounded top of the plunger. All these stop blocks are tightened up with handles instead of providing hexagon nuts, so that no separate spanners are required.

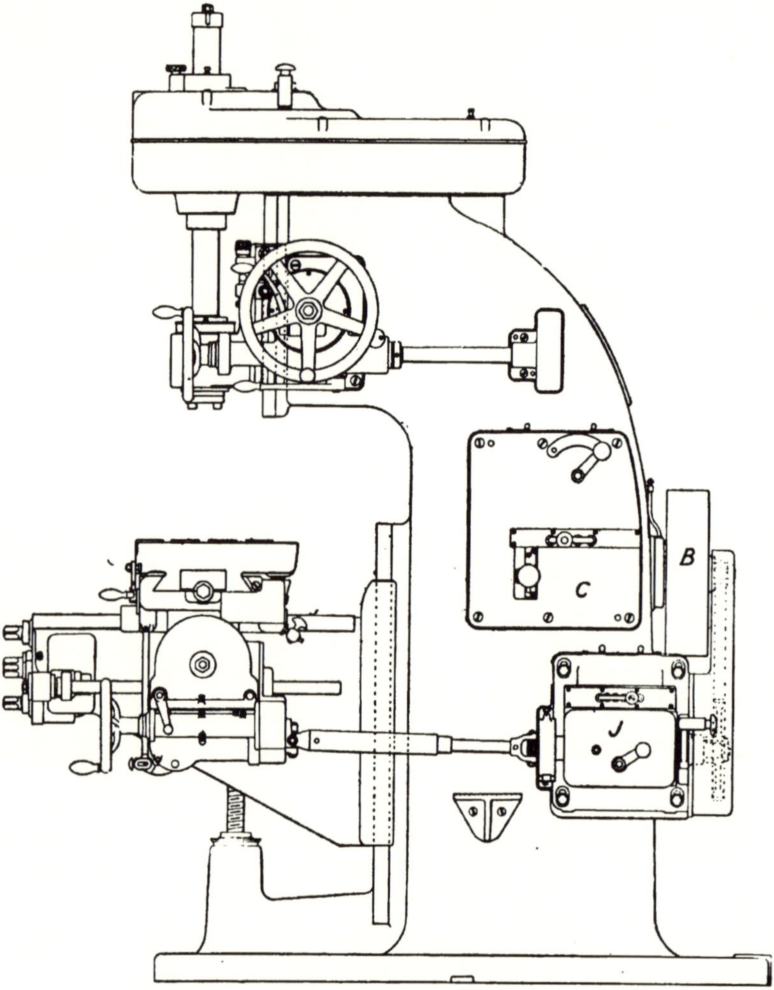

Fig. 98.—Side Elevation of Vertical Spindle Machine.
(B. & S. Manufacturing Co.)

Very complete arrangements are made for lubrication. There are the pump, tank, and feed pipes seen in the general views, Figs. 87 and 88, and the numerous waste-oil trays and lubricators shown

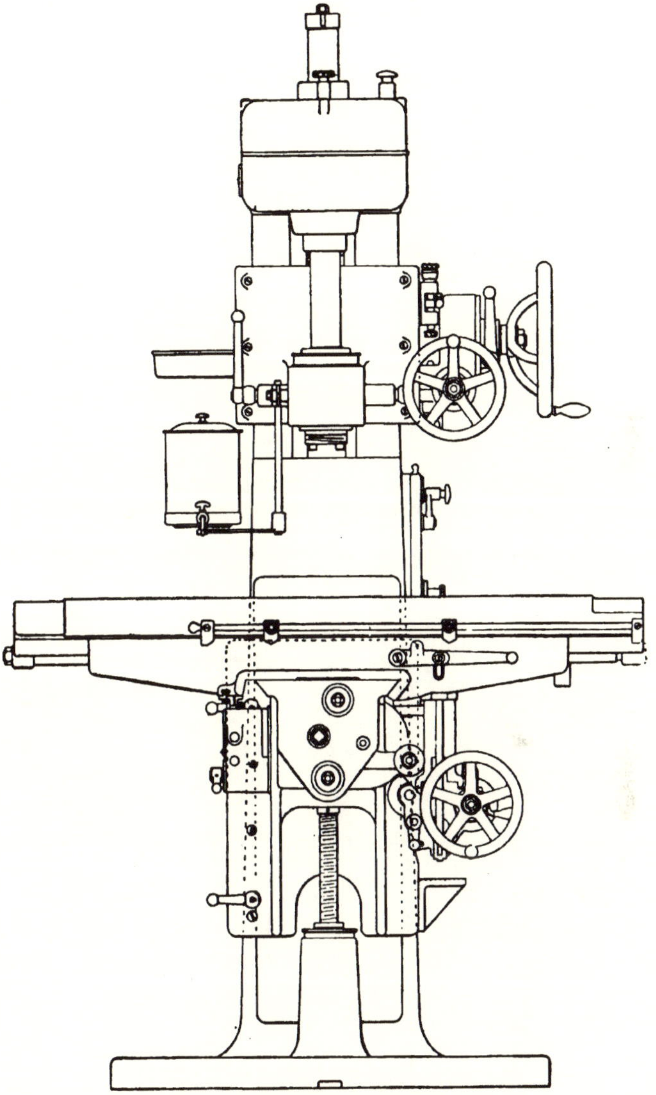

Fig. 99.—Front Elevation of Machine.

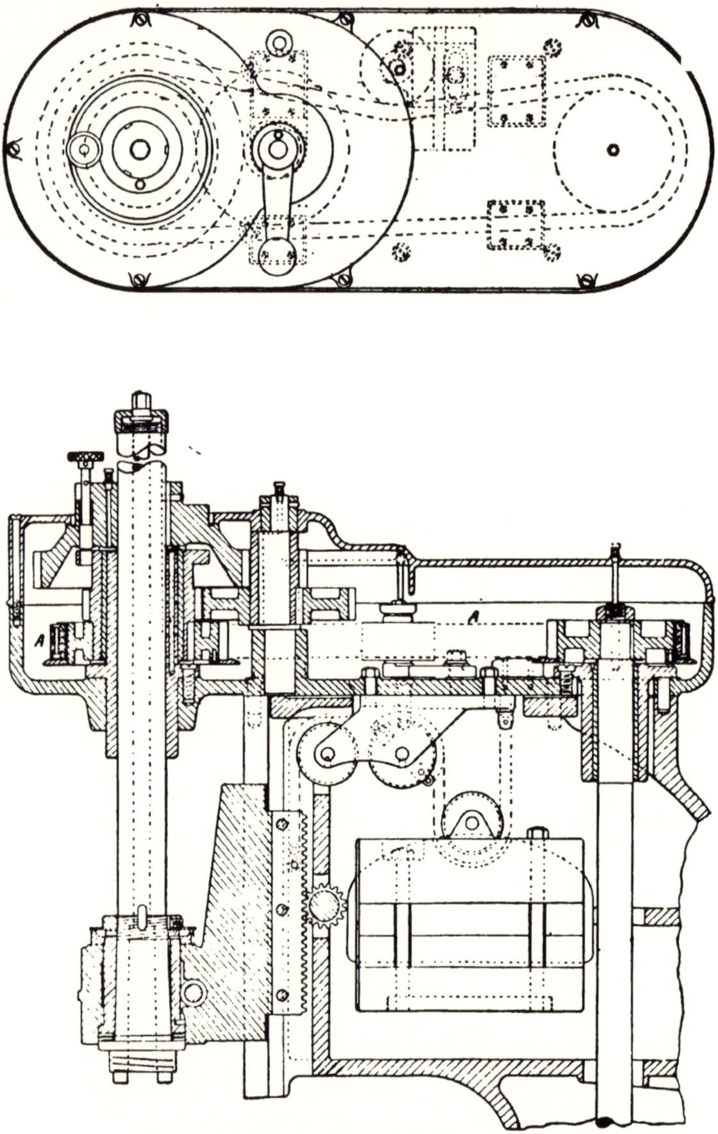

Fig. 100.—Plan of Head, and Sectional Elevation of Head.

in the details. Especially should those in Fig. 92 be observed, to be supplied through the doorway marked "Oil Gears," Fig. 93. Also

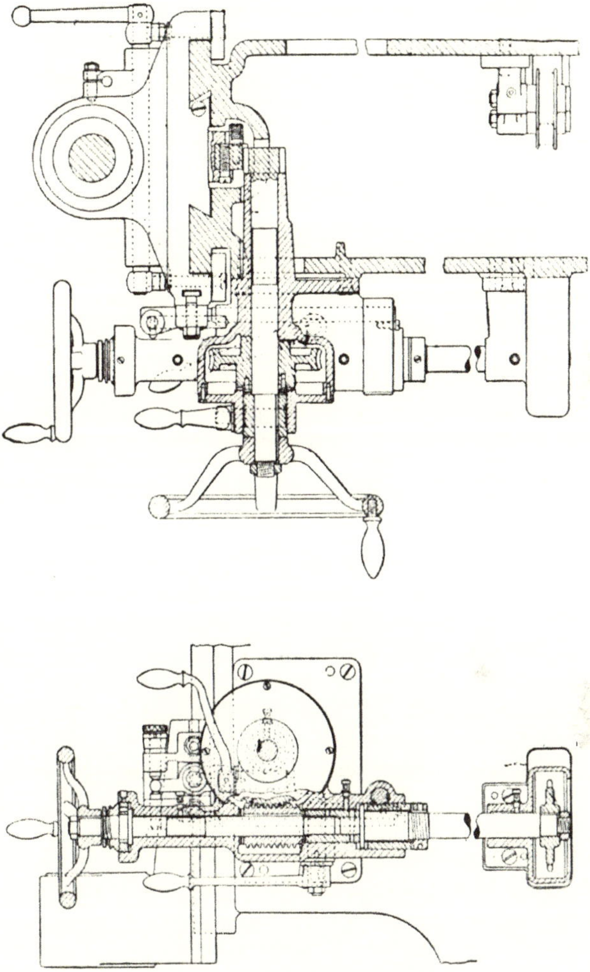

Fig. 101.—Details of Spindle Feed.

the taps for drawing off the lubricant from the trough around the table in Fig. 95. When the full system of piping is fitted, unions go in these holes, and allow the liquid to run down through the flexible pipes, shown in the general views, into the tank to be again pumped up.

The drawings of this beautiful machine will repay a more extended study than we have afforded it, many minor details being passed over without comment. A high-class machine tool such as this possesses an interest and fascination which increases with closer knowledge.

Figs. 98-103 illustrate the latest Brown & Sharpe vertical miller. Its principal interest centres in the feeds, which are derived from a shaft running at constant speed, instead of from stepped cones. The general outlines of the machine are shown in Figs. 98 and 99. It is of the elevating knee type, the vertical movement being 15 inches. The spindle has a range of vertical feed of 4 inches up or down. The latter makes from 17 to 354 revolutions per minute, and has fine hand feed, and quick return by separate hand wheels. The table travels 34 inches, and has a transverse movement of 13½ inches. The arrangements for feeding are to the right of Fig. 98, driving to the telescopic shaft.

Fig. 100 shows the head in plan view and in sectional elevation. The spindle, of crucible steel, runs in bronze bushings, the lower one of which has provision for adjustment. The spindle is hollow to receive a draw-through bolt for cutters, and the end is coned, threaded, and clutched. Its mass and that of its sliding bearing is counterbalanced, as shown in Fig. 100.

The mechanism of the spindle feed is shown in Fig. 101. The fine hand feed and quick return are effected by different hand wheels. The first is through worm gear, the second by a pinion gearing into the rack direct. The worm shaft is automatically driven from a sprocket.

The spindle is driven by a Renold chain A, Fig. 100, from a vertical shaft at the rear, which is connected by bevel gears with the driving mechanism at the rear of the pillar. The spindle is back geared in the ratio of 4·97 to 1. All these gears are enclosed (see Figs. 98, 99, and 100).

The drive from the countershaft (or motor) takes place to the pulley B, 14 inches diameter, Fig. 98, whence the mechanism of the gear box C is actuated. This is shown in detail in Fig. 102.

The shaft of B makes about 310 revolutions per minute. Between this and the shaft which is geared to the vertical shaft through the bevel wheels, there is an intermediate shaft D that carries

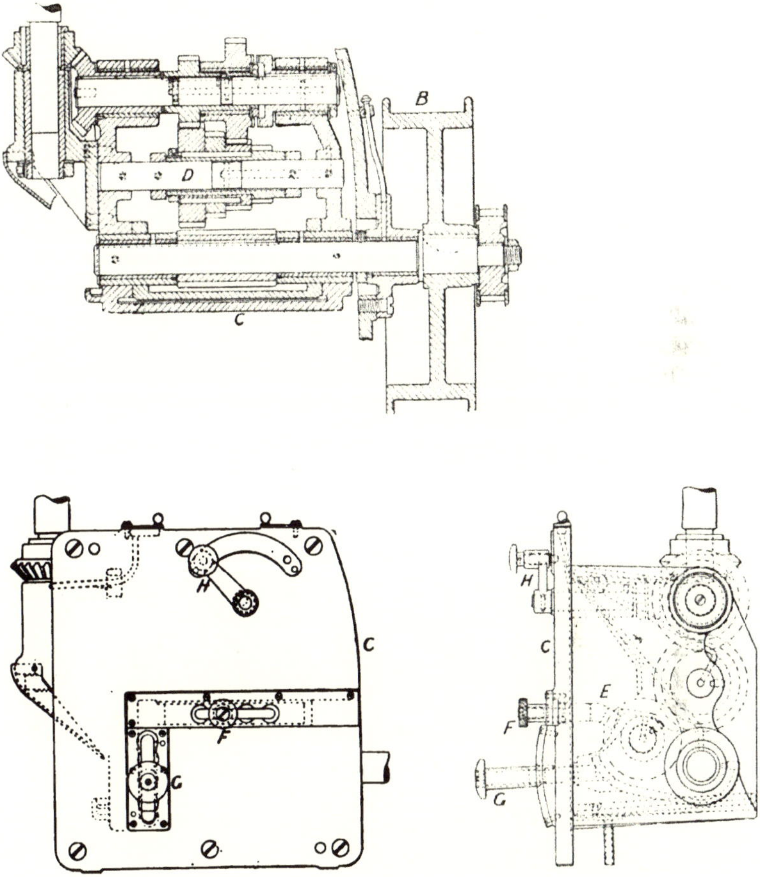

Fig. 102.—Details of Variable Feed Mechanism for driving Spindle.

four gears. Through the medium of an idler indicated at E, either one of these is engaged with a long pinion on the driving spindle. On another shaft are two gears constantly engaged with two of those in the cone of gears, and either of which may be engaged by the splined clutch seen to the right. The knobs F and lever G control the idler. F slides, and G unlocks and locks by downward and upward movements respectively. The position of the index slide for a given speed is set by bringing the knob F opposite the column of spindle speeds on the outside of the box.

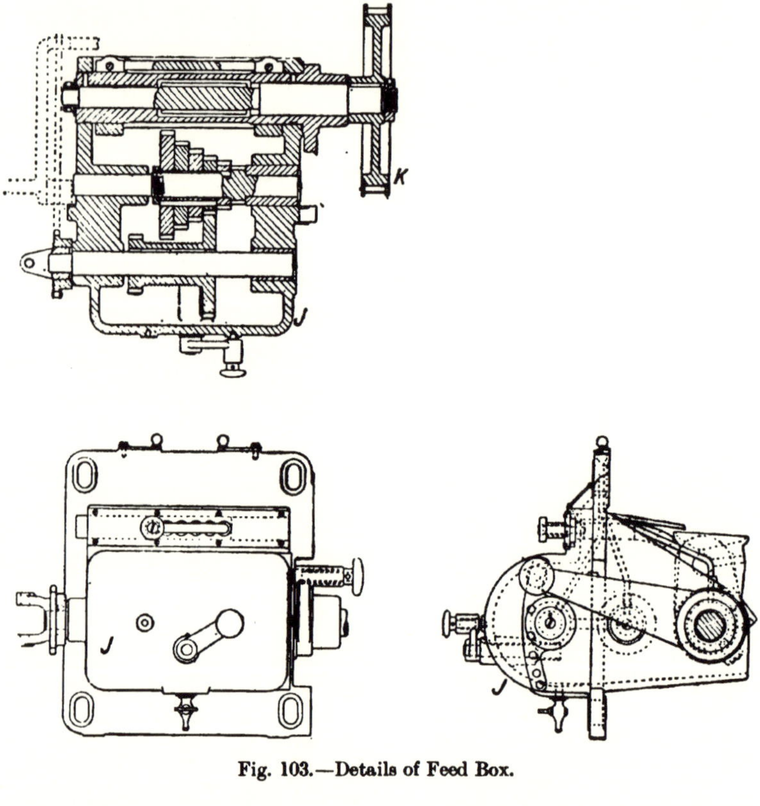

Fig. 103.—Details of Feed Box.

The crank handle H controls the two clutch gears above, giving two series of speeds. In this way 16 speeds are obtained through gear box, and back gears, ranging from 17 to 354 revolutions. The gear ratios range from 1 to 1, to 18 to 1.

The feed box is seen at J, Fig. 98, and its details in Fig. 103. The general design is similar to that for the spindle drive, though not quite identical. It is driven at a constant speed through a chain to the pulley K, thus rendering the table feeds independent

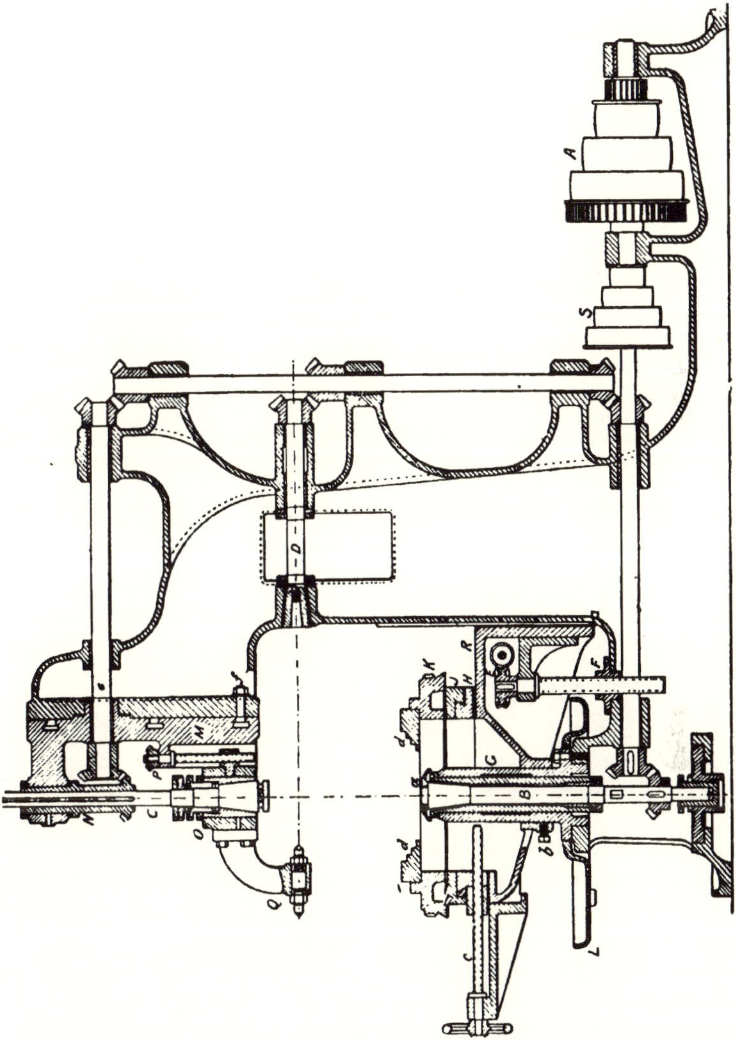

Fig. 104.—Vertical Spindle Machine.

of the spindle speeds. A cone of six gears is driven through an idler and thence to the telescopic shaft. A sliding knob and lever, seen in the box below, gives the feeds in inches per minute, ranging from 1½ to 17.

The table has, in addition to its automatic movements, a hand feed by the wheel seen below the table in Figs. 98 and 99. It has a quick return through a screw of coarse pitch, to effect which the feed worm is disengaged and a crank handle used.

There is a valuable class of machines, at present the speciality of one firm, that of Curd Nube, of Offenbach-on-Main, the leading feature of which is that the principal spindle comes up from below through the table. This feature alone is embodied in some of the machines by this firm, and there is consequently no over-hanging arm, and nothing above the table. But in another group, an upper vertical spindle is also included, and in another, a third spindle horizontally. In addition the head swivels for angle. Fig. 104 illustrates a machine of this complete type in vertical section. The spindles can be operated in unison or independently, and the upper spindle is capable of receiving vertical adjustments. The machine is capable of doing milling, drilling, and die cutting. The tables have compound movements. In machines which have no upper spindle, the table has provision for vertical adjustment.

The essential mechanism of the complete three-spindle machine in Fig. 104 may now be traced out. All three spindles are driven from the cones A, back geared, each spindle B, C, D, through its own set of mitre gears. The top bearing of B is protected with a cap *a*. The spindle fittings, and the method of taking up wear are similar in each, the lock nuts pulling against the coned bearing necks. The method of attaching the arbors is only indicated in the horizontal spindle, comprising a screw in addition to the taper of the shank.

The compound table is carried on the knee K, capable of vertical adjustment on slide faces at the front of the main framing. It is elevated and lowered by the worm gears E, actuating a vertical screw, threaded into the boss F, which is let into the machine framing. The knee also encircles a boss G, fitting by a turned check into a bored hole in the framing. It is clamped on G by the set screw *b*. On the knee the traverse slide H is moved to and from the pillar by the screw *c* and its hand wheel. The transverse slide J is moved at right angles to the direction of that of H by its screw and hand wheel, and the circular table

K is rotated on J by worm gear, the wheel being cut solidly around the edge of the table. K carries stepped chucking strips *d, d*, which have setting-up and clamping screws (not indicated), and by means of which dies or other pieces with parallel or with tapered edges are gripped. It is thus seen that though the bottom spindle B has no vertical movement, the table can be adjusted for height with great precision, and within a large range, and clamped securely, so being adaptable for work of different depths, and for coarse and fine feeding. Also that cuttings fall away down into a hollow space in the knee, and together with waste lubricant are received in the tray L. Any other chips and lubricant are received by the channel that is formed all round the foot of the main framing.

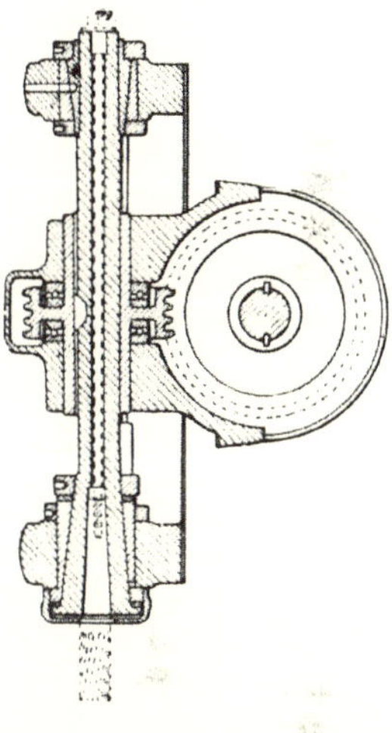
Fig. 105.—A Pratt & Whitney Spindle.

The vertical spindle C is carried by the swivel plate M, which pivots around the axis of the driving shaft *e*. M has an index or zero mark on its flanges by which the spindle is set perpendicularly, or to an angle, to right and left, with graduations on the corresponding flanges of the main framing, and is clamped by bolts, one of which is seen at *f*. The spindle C can slide in the sleeve N, thus accommodating itself to the vertical adjustments of the sliding bearing O. The latter moves on flat slides on the swivel head M, when operated by the mitre wheels P, and screw, the wheels being turned by a hand wheel standing out to the front (not shown). O receives an overhanging arm Q, with a centre for the arbors held in the horizontal spindle D. S is a pulley which supplies self-acting feeds to the table when required.

This is a brief description of one of a great group of machines which possess a very wide range of utility.

One method of elevating the table in one of the types of machines by this firm is this: the support to the tables has a long cylindrical boss, encircling a long boss that forms a prolongation of the base. A large hand wheel is threaded to fit this, and serves to elevate the table slides bodily.

Fig. 105 illustrates the spindle of one of the double-spindle profiling machines by the Pratt & Whitney Company. A horizontal shaft, splined, carries a spiral gear which engages with another

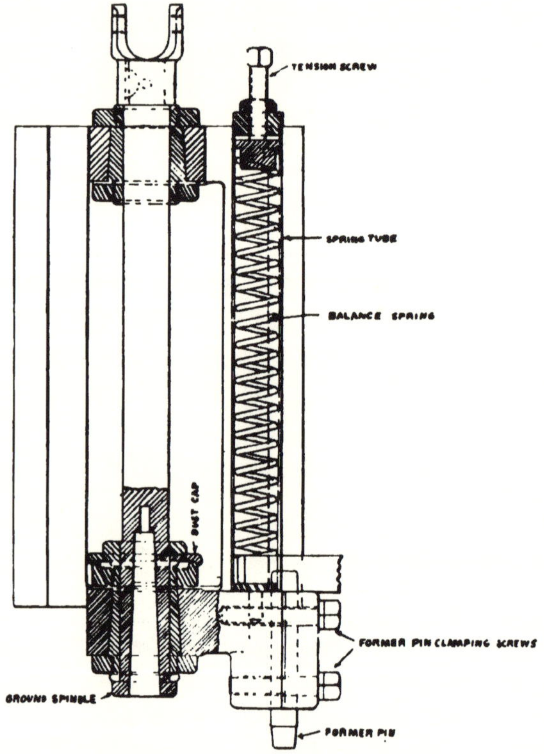

Fig. 106.—Spindle of Garvin Profiling Machine.

spiral gear, encircling the cutter spindle by a long boss, and keyed to it. The pressure on the spindle, and its own weight are taken on ball thrusts. Wear is taken up on conical necks with ring nuts. The cutters fit by taper shanks, and are held in by a long bolt that passes up through the hollow spindle. The nose is protected by a cap.

The Garvin spring-balanced slide is shown in Fig. 106, in which the spindle, driven at the top by a universal joint, runs in adjustable bearings in a casting, which carries also the former pin. The balancing spring is contained within a tube, and is provided with an adjusting screw.

The Work of the Vertical Spindle Machines.—Excepting work involving the employment of the spiral head and swivel table, these machines include everything, plane and curved, that comes within the range of their individual capacity, including face and edge milling. They are equally adapted for both of the latter methods of cutting. But the machines are eminently suitable for curved work, using edge mills, and the range of radius covered is large. Combinations of curves with straight parts are also as readily done, by combining the movements of the linear slides with those of the circular table. Work that is bored can be chucked by the bore, through the arbors which are fitted to the centre of the table, as in slotting machines. Faces of different depths can be done with face mills, which is convenient in cutting recesses such as those in smiths' dies for stampings.

The Work of the Profiling Machines.—The general mechanism of these machines has been described, pages 96 to 104. The work done consists in making and attaching a suitable former to the table of the machine, and adjusting the work to the milling cutter, and both in relation to the former pin. In some cases provision is made for adjusting the latter, but generally there is none, excepting sometimes the choice of two holes, in either of which the pin may be clamped. The pin is tapered, which permits, by its vertical adjustment, of deepening the successive cuts.

CHAPTER V.

PLANO-MILLERS OR SLABBING MACHINES.

Plano-Millers or Slabbing Machines—Their Characteristics—Details of Spindles—Horizontal—Vertical—Rotary Planers or Ending Machines.

Plano-Millers or Slabbing Machines.—This is a name that seems most appropriate to designate that large and growing group of machines which is built on the model of the common planing machine, with bed, table, housings, and cross rail carrying spindle heads. These are the most obvious rivals to the planing machines. They differ from planers in the slower table feeds, and in the character of the vertical feeds imparted to the milling cutters. But the plano-millers have gone beyond this design, for many machines include both horizontal and vertical spindles, and some with angular movements, others also have profiling attachments, and many include circular tables on the reciprocating ones. Besides this, some machines have two tables moving side by side, and these again may be run independently, or coupled for simultaneous movement, to carry wide work. Then there are numerous machines convertible into the open side type, and special forms of these.

The principal points which should characterise a good milling machine of the planer type, Fig. 107, are, besides massiveness in framework and in spindles, and ample bearing proportions: provision for taking up wear, balanced slides or heads (when these are of considerable weight), easy, accurate, and smooth movement to the table; a wide range of table speeds and cutter feeds, provision for rapid movement of the table in each direction for setting work, quick movement of the heads for setting spindles, fine vertical adjustments for the cross rail, fine adjustments for the table and heads, self-acting feeds to the table, with automatic knock-out; provision for lubrication by pump, an oil trough round the table, ample belt and gearing power, and convenient location of handles for the various movements. In proportion as a given machine embodies these general characteristics may its value be estimated. All machines do not include the whole of these: some are deficient in most of them; some are very well designed in regard to certain details, while deficient in others.

In selecting a slabbing machine regard also should be had to the class of work which has to be mainly done. If both faces and edges on the same piece of work have to be tooled, then, having a horizontal arbor only, the work will have to be re-set for each separate face. But if, in addition, there is a vertical arbor, then two faces can be tooled at one setting. If there are three or four arbors—which is the case in some machines—the range of duty is correspondingly increased.

The number of firms who make small milling machines of excellent design and workmanship is greater than that of those who construct heavy ones. One reason lies in the fact that fewer shops are equipped for the building of heavy machine tools than of light ones. But the principal reason is that the practice of heavy milling is of much more recent development than light milling. The economy of heavy milling *versus* planing is as yet a matter open to question in the minds of some engineers. The planer type of machine is, however, becoming more and more adapted to the work of heavy milling, and cutters are made more capable of taking heavy feeds.

As remarked in a previous page, the early uses of milling were all confined to the tooling of small articles, and not at all to service in general machine shops, nor for heavy work. Of late years that has been changing, and a good deal of heavy milling is now done on the slabbing machines, both with edge mills and with face mills having inserted teeth. During this development lessons have been learned, the results of which have been utilised in later designs, so removing the objections and prejudices to heavy milling, which have been due to want of knowledge of proper conditions and necessary limitations. The stresses due to the use of wide cutters with continuous edges being much greater than those due to the use of single-edged tools are the cause of the greatest difficulties in milling considerable widths. Though these are diminished by the spiral grooving of cutters and by staggered teeth, true work, when of considerable width,

Fig. 107.—Plano-Miller.

can only be done by the use of stiff machines, stiff arbors (by supporting both ends of the arbor in long bearings), or by both devices in conjunction, and by taking shallow cuts and slow feeds. Without these precautions, milling which is at once broad and accurate is impracticable. And so we find that the best machines to-day are designed and constructed of very massive proportions. Yet the best are none too stiff now for much of the duty imposed upon them.

Of the planer types of milling machines there are various modifications. The first variation occurs in machines in which one upright is made either extensible or removable, to receive work wider than the normal capacity will take. Another occurs in the entire removal of one upright, making a permanent form of open-sided machine. Each of these general designs is constructed with variations in detail.

The four-spindle machines are specially designed to operate on two faces, vertical and horizontal respectively, of two pieces of work at one time. There are two sliding on the cross rail, and

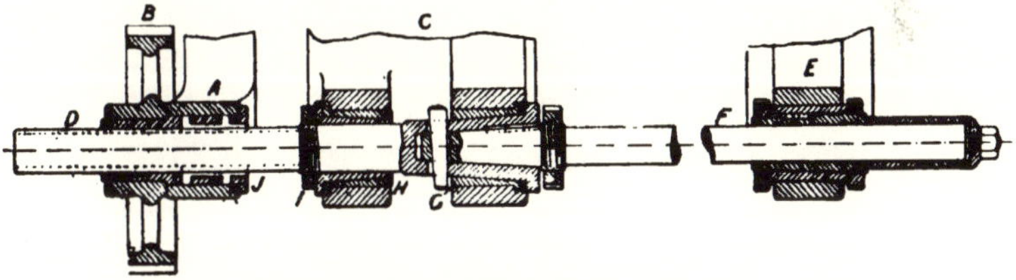

Fig. 108.—Ingersoll Spindle (Horizontal).

two on the uprights. All are driven from one cone shaft, and all have independent adjustment.

The spindles of the Ingersoll machines are constructed in the way shown in Fig. 108, being for the 36-inch size. In these figures A is a fixed bearing on the cross rail, B the wheel which drives the spindle, C is the movable bearing on the cross rail, and D the parallel portion of the spindle, which slides in the boss of A, as C is adjusted endwise. The arbor support E also can be slid endwise on the cross rail and over the arbor F. It will be seen that the spindle is fitted with double reverse cones. The front cone G is kept pulled back to its bearing by the adjustment of the hinder cone H, tightened by the lock nut I. The cone, which encircles the spindle, is of cast iron. Its bearing is of phosphor bronze. The wheel B drives the spindle, while permitting its parallel end D to slide endwise through it in the sleeve extension J.

This sleeve runs in the bearing of A, and is prevented from moving endwise in either direction, the spindle D moving along with C. A couple of keys are fitted in the bush. These slide in the key grooves in D, and rotate D in any position covered by the range of the key groove. The inner end of the arbor fits the tapered hole of the spindle. The outer end F runs in a parallel bearing in the bracket E, and wear is taken up by the lock nuts and cone.

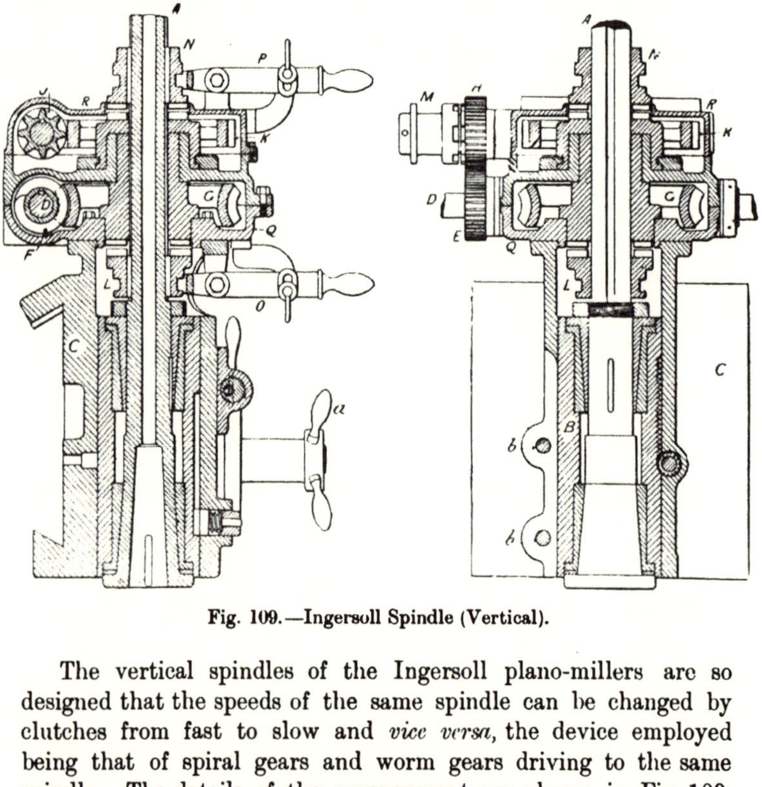

Fig. 109.—Ingersoll Spindle (Vertical).

The vertical spindles of the Ingersoll plano-millers are so designed that the speeds of the same spindle can be changed by clutches from fast to slow and *vice versa*, the device employed being that of spiral gears and worm gears driving to the same spindle. The details of the arrangement are shown in Fig. 109, which represents two vertical sections taken in planes at right angles with each other.

In these figures the hollow spindle A has its bearings in a sleeve B that slides vertically in a boss cast solidly with the saddle plate C, which traverses on the cross rail. The sleeve is adjusted by the pinion and rack shown, operated by the cross handles *a*, and is clamped by the screws *b*, *b*.

The rotary movements of the spindle A are derived in the first place from the horizontal feed rod D, splined to drive the spur wheel E and the worm F. F engages with its worm G, and E drives the spiral gears through H; H being on the end of the spindle that carries the small spiral wheel J engaging with the large wheel K. Wheels G, H, and K are all loose on their spindles, K and G fitting with sleeves, and are therefore put into engagement by the sliding clutches L, M, and N respectively. L and N are slid along by the locking levers O, P. The worm and spiral gears are enclosed by the casings Q and R.

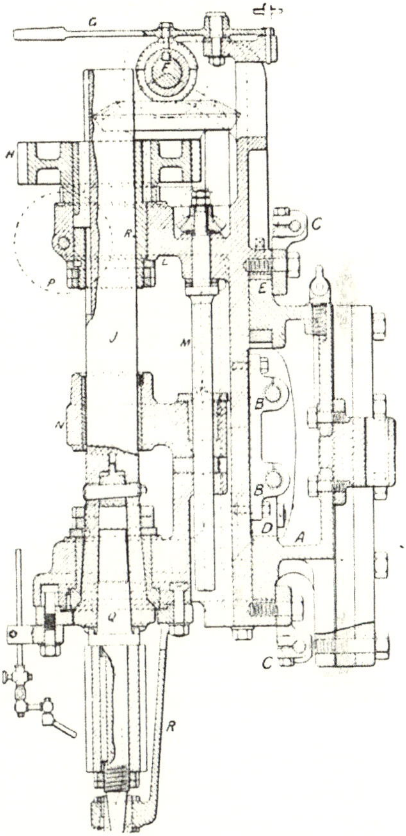

Fig. 110.—Hetherington Spindle (Vertical).

Fig. 110 gives a vertical section of the spindle details of a massive double-headed machine of the planer type by Messrs John Hetherington & Sons Ltd.

The cross slide is seen at A enclosing the two feed screws B, B, one for each head. C, C show operating dogs, adjustable along the feed rod, by which, through clutches and levers, the screws B, B are tripped. The rollers by which the friction of the carriage is lessened when profiling, run on the lower flange of the cross slide, one being shown at D. The take-up strip E, of gun-metal, is on the top flange. The milling spindle on each head derives its motion from the horizontal shaft F. A nest of bevel wheels with spurs connect shaft and spindle, and a clutch, moved to right or left by the lever G between the bevels on the shaft F, effects reversal. The driving spur pinion is on the crown bevel. The driven spur H drives the spindle J through an intermediate sleeve R, which runs in the bearing L. The object of this device is to relieve the spindle of strain due to the pull of the belt. If the wheel H overhanging its bearing had no such aid, the pressure of driving would cause the spindle to wear the bearing to one side. M is the vertical feed screw operated by bevel wheels above, driven from a horizontal shaft. The screw runs in a gun-metal bush in the bottom spindle bearing N. Wear is taken up by lock nuts P and similar nuts below. The arbor Q fits with a taper of 1 in 20. It is secured with the tang and screwed cottar seen. Clutch jaws engage the spindle end to give a positive drive. A support or stay to the lower end of the arbor is included, seen at R. This carries a gun-metal bush for the arbor end, tapered 1 in 6, and the bush has adjustment for wear.

The three or four-spindle milling machines of the planer type are admirable tools for a large class of work hitherto done on the planer. They ensure the correct tooling of three faces at right angles without re-setting the work, and using face cutters with inserted teeth, they remove material rapidly. If sufficient care is taken they will also effect fine finishing cuts accurate enough for all the ordinary run of engineers' work, such as faces against which attachments are to be bolted, the feet of brackets, the ends of distance pieces, and specially that class of work in which ends and edges or surfaces and edges have to stand at right angles. They are also admirably adapted for tooling two independent sets of work simultaneously when there are two vertical spindle heads and two horizontal ones. The utilities of these machines, though recognised more than they were three or four years ago, are not appreciated as they might be.

The space available for work set on the plano-milling machine is not so large as that on the beds of the rotary planers, Fig. 111, with traversing heads. The space is limited in the first by the capacity between heads. In the latter the work may cover the area of the

Fig. 111.—Rotary Planer.

table, and extend much beyond it if necessary. The difference is analogous to that which exists between the limitations of the ordinary planing machine and the large dimensions which can be got upon the beds of the vertical planers and of the edge planers. The classes of work, therefore, for which the two machines are specially serviceable are different. Each can be used for fine tooling. But the ending machines are frequently employed for jobs which would otherwise be done on the plate-edge planer, or with a cold saw, or in the lathe, or with a vertical planer—such work as squaring the ends of beams and girders, facing column ends, ends of standards, and engine and machine framing of various kinds, besides tooling pieces arranged in series on the bed.

CHAPTER VI.

SPECIAL MACHINES.

Special Machines for Gear Cutting, &c.—For Spur and Bevel Gears—For Worm-thread Milling—For Hobbing Worm Threads—For Fluting Twist Drills—Three-Spindle Machines—Cam-cutting Attachments—Profiling Mechanism—Milling Attachment for Planer—Machine for Elliptical Holes.

Special Machines for Gear Cutting, &c.—Figs. 112-119 illustrate a machine by Brown & Sharpe for cutting both spur and bevel gears. In the main the machine is built on the pattern of the spur gear cutters by this firm, with the difference that the cutter slide is made to angle. This example is selected as representing a modern high-class fully automatic machine, which embodies the following valuable features:—

Stiffness of build, a large number of changes of speed and feed, to be noticed presently. Protection of the working parts. Support afforded to the cutter spindle. A high speed of positive indexing, which, with the return of the cutter slide is independent of the speed and feed of the cutter. The elevation and lowering of the work spindle slide by power in the largest machines. The drawings, with the following brief remarks, will be readily understood.

Fig. 113 is a rear elevation of the machine, Fig. 114 a right-hand end elevation, Fig. 115 a plan view. A, the base, is fitted as a cabinet, and B, the pillar, is fitted with shelves for gears or tools. A carries the cutter slide C, D; D being adjustable on C to any angle to 90°, by means of a quadrant rack, so that it is suitable either for cutting spurs or bevels. The slide C has longitudinal adjustment by hand through pinion and rack, and power feeds and quick return operated by gears enclosed. The motion is communicated at any angle of the slide D, through the splined shaft E, by means of bevel gears at both ends, enclosed. The pulley F, driven from the counter-shaft, drives the cutter spindle, and the pulley G, the cutter slide, through gears and a quick pitched screw, so making the movements of slide and spindle independent of each other.

The cutter spindle has ten changes of speed, ranging from 30

Fig. 112.—Automatic Spur and Bevel Gear Cutter.
(B. & S. Manufacturing Co.)

to 163 revolutions per minute obtained through change gears. Fig. 116 gives a section through the spindle and its bearings, with the details of the method of adjusting it.

The cutter has sixteen changes of feed, ranging from 0·012 to

0·235 inches per revolution of spindle, in geometrical progression. These are obtained through change gears enclosed in the box H. When cutting bevel gears there is a cross slide which can be moved to bring the cutter to either side of the centre. The mechanism for this movement is seen at J, Fig. 115, while the vernier is seen in Fig. 118.

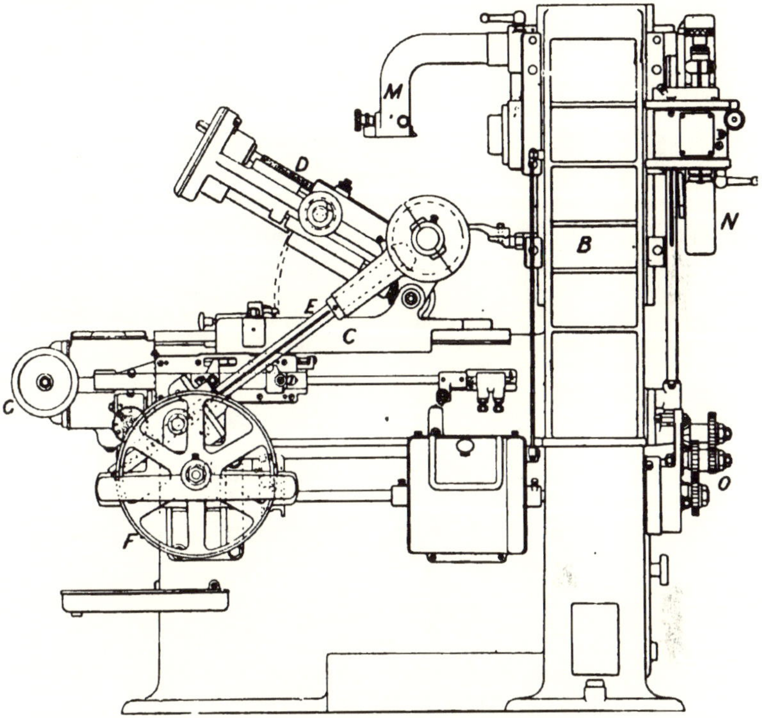

Fig. 113.—Rear Elevation of Gear Cutting Machine.

The work spindle head K is adjusted by the hand wheel, Fig. 115. The thrust of the elevating screw is taken by ball bearings. A graduated dial reads to thousandths of an inch. The spindle is hollow, with a $1\frac{1}{4}$-inch hole, and tapered at the front to No. 12 taper. It is threaded to receive a face plate or other fixture. An overhanging arm M affords support to the outer end of the arbor. Large gears are supported by a rest behind the rim opposite the cutter. The worm wheel, enclosed in the casing N, is driven by a vertical worm, and the indexing change gears provide for cutting all numbers of teeth from 12 to 50, and all numbers from 50 to 400, excepting prime numbers and their multiples. The indexing mechanism is shown in Fig. 117. Fig. 119 is an indicator for setting the cutter.

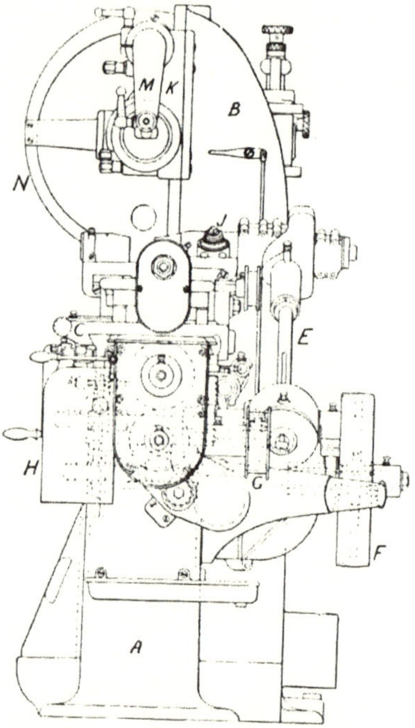

Fig. 114.—Right-hand End Elevation of Machine.

Worm-thread Milling.—A worm-thread milling machine, by J. E. Reinecker, is shown by the accompanying Figs. 120 and 121. There is no doubt that the milling of worm threads, though a comparatively recent innovation, will in time displace the single-cutting lathe tool for that class of work. The cutter, of rack section, gives no trouble with regard to clearing itself. The present demand for multiple-threaded worms is the opportunity for the development of machines of this class, which are now made in Great Britain, the United States, and Germany.

As the cutter in the Reinecker machine is fed longitudinally in relation to the worm blank, while the worm is rotated, this involves two distinct sets of mechanism, each with its change gears. Figs. 120 and 121 show the arrangements of the machine as a whole.

The arbor, indicated at *a* is carried in the headstock A and tailstock B. The spindle is made hollow, in order to take the axles of worms, when in one piece with their worms, so that they can be gripped and supported close to the threads. The cutter is held on the arbor *b*, the bearings of which are carried on the plate C, adjustable for angle on the face of the head D. The blank is rotated from the cones E, through the two sets of worm gears F,

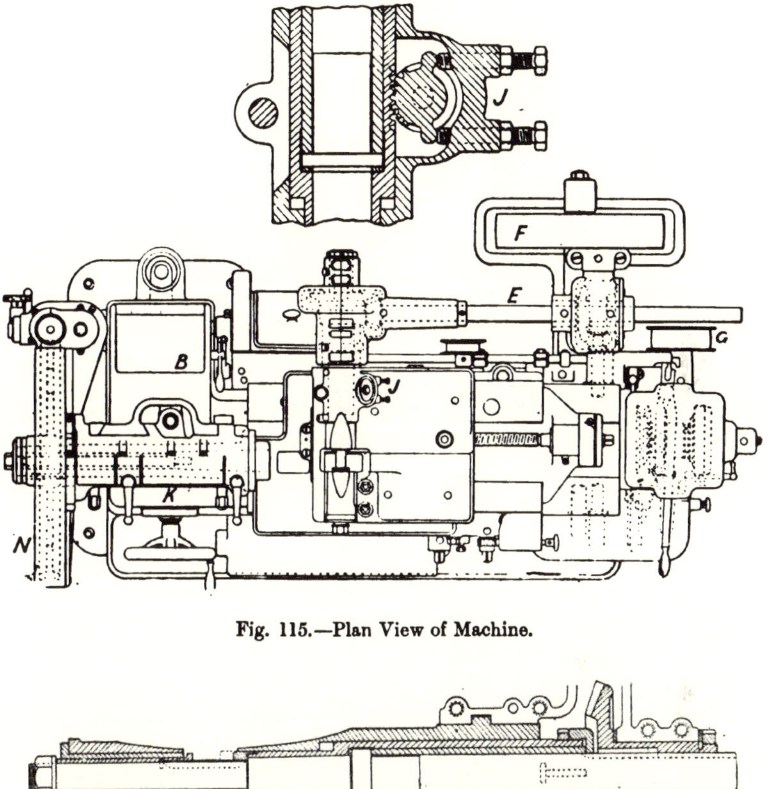

Fig. 115.—Plan View of Machine.

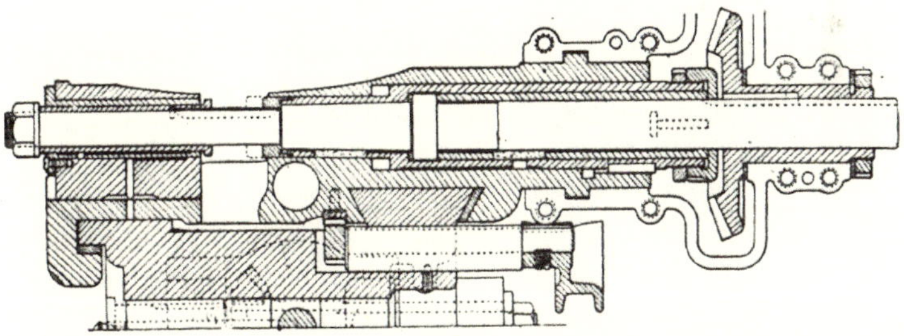

Fig. 116.—Section through Cutter Spindle, showing Method of Adjustment.

and G. The shaft *c*, that carries the worm that drives G, on the headstock spindle, also actuates the set of change gears H, whence the correct longitudinal feed is imparted to the head D, which carries the cutter arbor. The due proportion or ratio between the diameter and the pitch of any worm being milled is thus obtained by changes in the gears H, the worm drive being constant from the

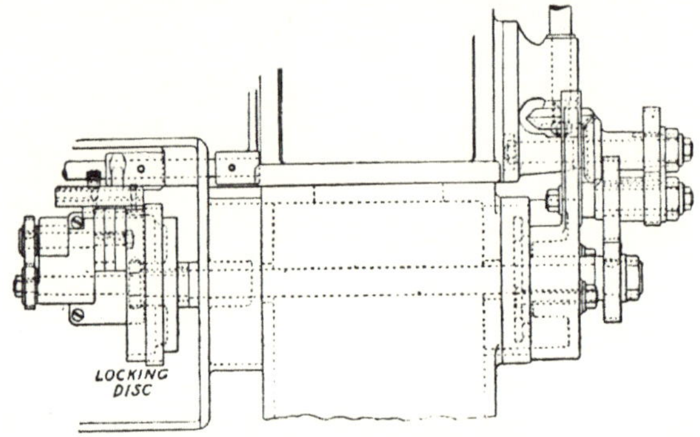

Fig. 117.—Indexing Mechanism.

stepped cones E. Changing the belt on the steps of E does not, of course, affect the relations of the worm gears and change wheels.

The rotation of the milling cutter is effected by an independent set of mechanism, derived from the cones J. The course of the motion can be traced from J, through bevels and spurs to the last spurs, K and L. K, it will be noticed, has the axis of its shaft central with the swivel plate C, and L runs in a bearing in a boss projecting inwards from the plate C, so remaining in gear with K at all angular positions of the cutter arbor *b*; L drives a spiral gear *d* through the plate, which engaging with another *e*, on the arbor *b*, rotates the latter at four variable speeds obtained from the cones.

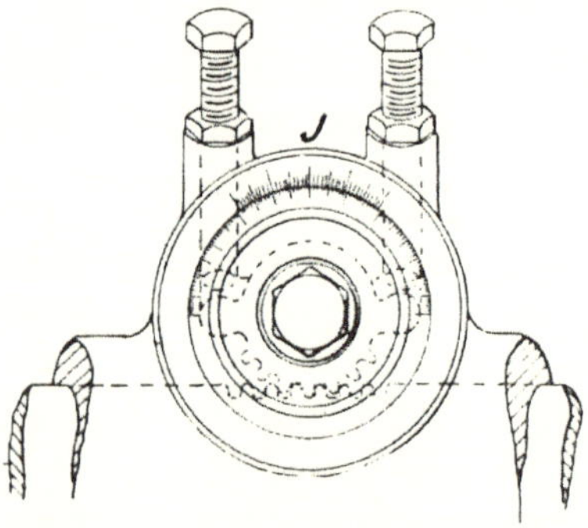

Fig. 118.—Vernier and Screws for adjusting Cutter Spindle.

There is an indexing device shown at M, over the headstock

spindle, for obtaining settings for multiple-threaded worms. This has been abandoned on later machines in favour of direct indexing with holes and index peg.

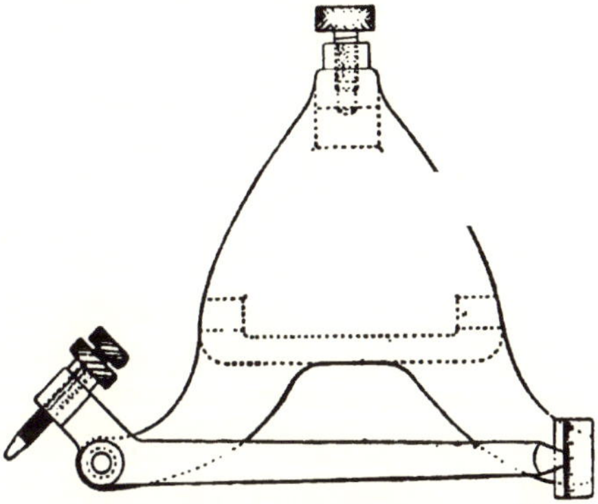

Fig. 119.—Indicator.

Hobbing Worm Gears. — A remarkable machine, Figs. 122 and 123, by the same firm, for hobbing worm gears by the method just described, also cuts both spur and spiral gears either with a hob, or with an ordinary cutter. Its movements are auto-

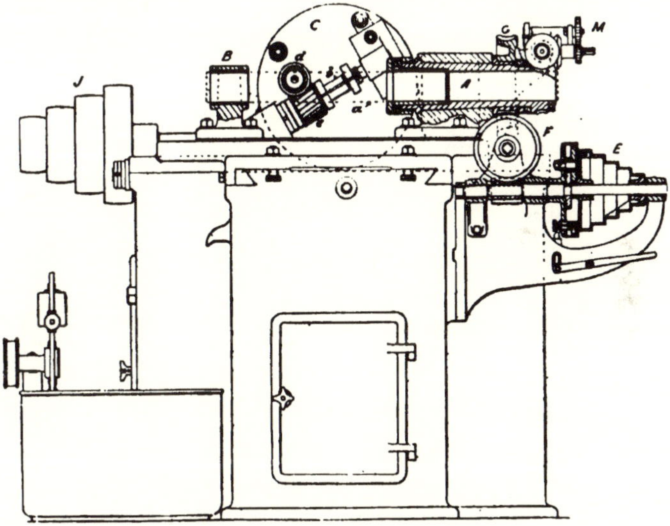

Fig. 120.—Worm-thread Milling Machine.

matic, and the results theoretically as well as practically perfect. The worm wheel blank to be hobbed is put on the arbor A, and the hob (which is fed in endwise) goes on the arbor B. The mechanism is designed to effect the relative movements required. The distance from the centre of the hob to the centre of the blank remains unchanged from start to finish of the operations. The rotation of the hob and that of the blank are so proportioned as to be exactly the same as those of the wheel and its worm. The longitudinal feed may be varied, and does not affect the pitch, or the diameter, and number of teeth in the worm and worm wheel. The driving mechanism is actuated from the

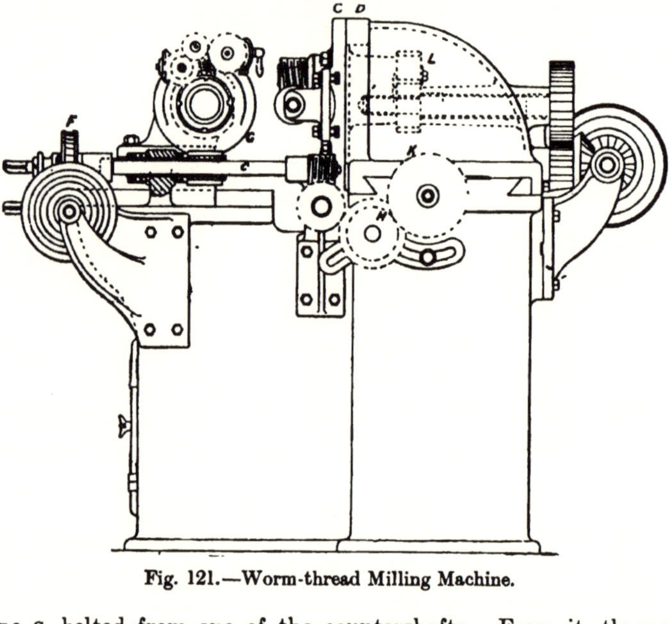

Fig. 121.—Worm-thread Milling Machine.

cone C, belted from one of the countershafts. From it, through the change speed gears D, the vertical splined shaft is actuated, and from this the rotation of the hob is effected by the pinion F, and wheel G, while the rotation of the master wheel H, and from it the blank, is imparted by a set of change gears (not indicated, but situated behind wheels F and G). The mechanism of this rotational movement is carried through the centre of the carriage to change gears K, at the opposite end of the table. The feed, or longitudinal travel of the table J, is derived from the feed cones L, driven from a separate countershaft. From L, the vertical feed rod M is driven through the worm and mitre gears seen in the figure, at variable rates, and thence the horizontal feed shaft N. The latter operates two sets of

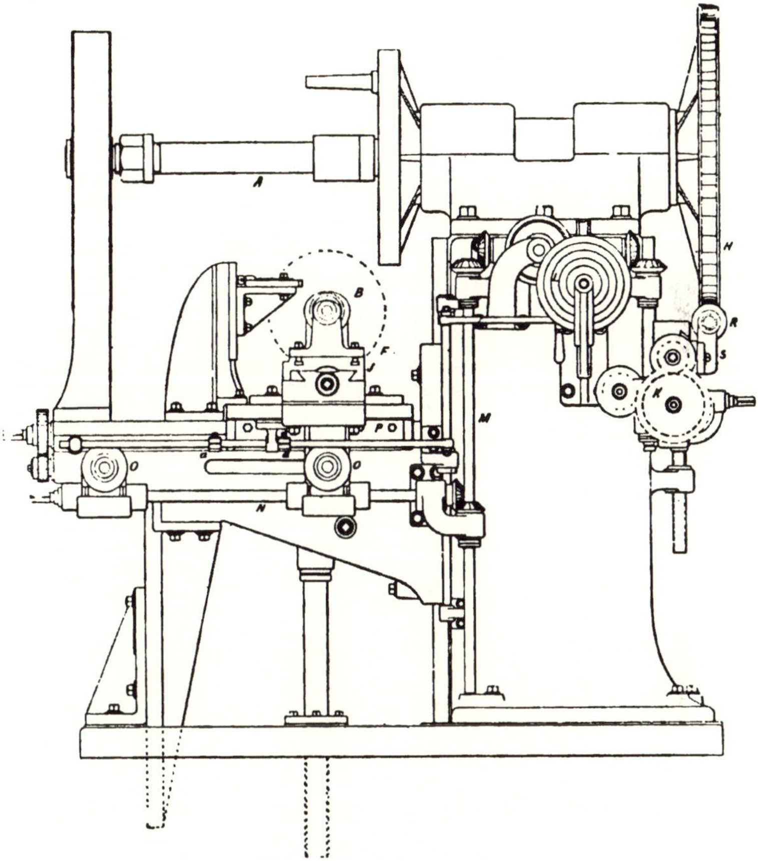

Fig. 122.—Universal Gear Cutter.

worm gears O, O, one for longitudinal traverse of the table J, the other for the cross traverse of the carriage P. Both sets of worm gears are clutched for throwing out and in. These movements are capable of reversal through the dogs *a*, *a* on the horizontal rod situated above the worm shaft N, which through various levers actuates the clutch of the bevel gears Q. The object of the change gears K, is to add the proper rate of rotation to the blank on the arbor A, to compensate for the forward motion of the

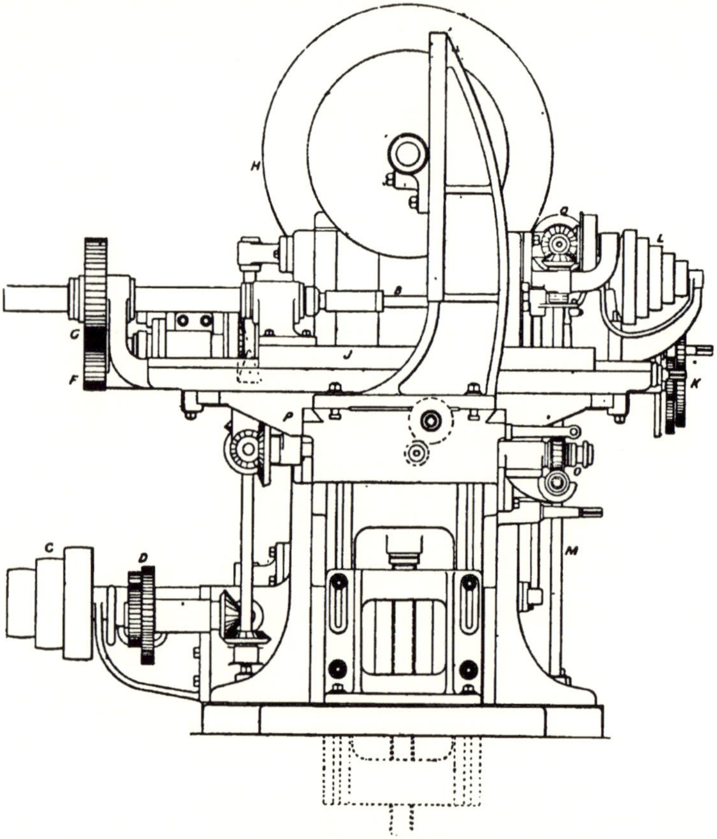

Fig. 123.—Universal Gear Cutter.

hob, which is the most marked peculiarity of this machine, necessary because the hob is fed at a tangent to the blank, and being so fed, the worm R, which drives the master wheel H, must be fed similarly, with which object its bearings are carried on the slide S. The ratios of these movements are the same as those existing between the hob and the wheel which it cuts, and the master wheel and its worm. The rates of rotation of the hob and worm are similarly proportioned.

For cutting spiral gears, the slide J is mounted on a swivel table, a special milling attachment being added. The dividing is done by hand by means of a dividing plate and change wheels. For cutting spurs, the carriage P is traversed towards and from the column.

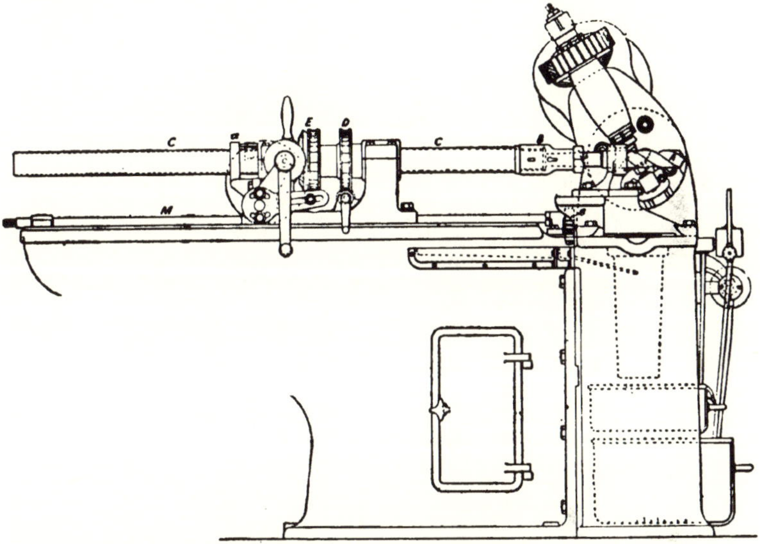

Fig. 124.—Flute Milling Machine.

Fluting Twist Drills.—Though the flutes of twist drills can be cut on a universal machine, only one flute can be done at a time. A special Reinecker machine, designed for cutting both flutes simultaneously, is shown by the illustrations. In it the drill is fed forward and rotated between cutters, the centres of which move apart to give the increase of thickness from point to shank that is necessary to strength. Relieving as well as fluting is done. The machine shown is the larger size of two made, its maximum capacity being a 4-inch drill, 40 inches long. The following is a description of the principal elements of the mechanism :—

Fig. 124 illustrates the machine in side view parallel with the drill. Fig. 125 is a front-end view looking towards the drill. The bed is made in two portions, standing at right angles and bolted together, one part carrying the drill and feeding arrangements, the other the spindle head.

The drill blank A is carried in a chuck B, which may have either a Morse taper, as shown, or a straight shank, and it is fed forward and rotated by a long screw C, splined to permit of its

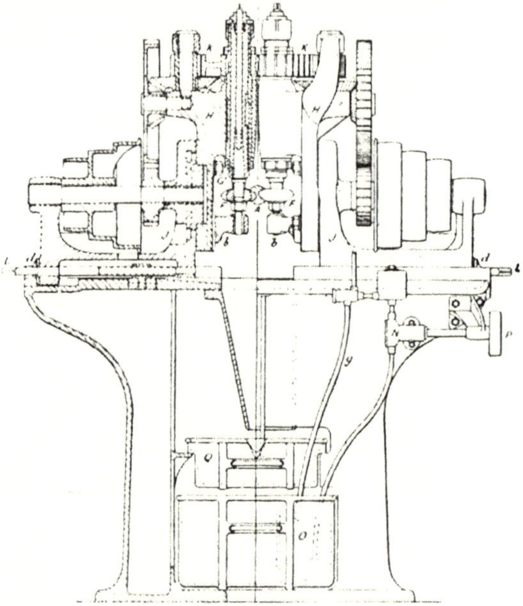

Fig. 125.—Flute Milling Machine.

endlong movement through its operating gears. The worm gear D feeds, and E rotates, and the rates of both are controlled by the change gears, indicated by the dotted circles, which establish the desired relation between the pitch of the feed screw C and the lead of the spiral of the twist drill. Rough adjustments of the forward and the rotary motions can be effected by the handles seen. A stop *a* can be set to gauge the termination of the cut.

The fluting cutters are seen at F, F. They are carried on arbors which fit by taper shanks in the spindles G, carried in long sleeves in the head H. The tail supports *b*, *b* are fitted in slide ways to permit of their movement and setting for removal and insertion of cutters. The swivel heads are adjusted for angle on the headstocks J, J, on which they are set with tee-headed bolts. The angle, of course, varies with drills of different diameters.

The spindles G have a small amount of vertical adjustment on a feather seen in the section, and they are driven on the sleeve, the spur wheel being keyed thereon to avoid direct pull and wear on the spindle. The drive to K is readily seen in Fig. 125, coming from a spur wheel on the same spindle as the stepped cones, thence through bevels to the spur that gears with K. The adjustments of the heads J, J towards or from the drill blank are effected by the hand-operated screws L, L, working through nuts in the base, the rectilinear movements being controlled by vee'd edges. Divisions on plates on the base of each head permit of making fine adjustments, and micrometric divisions are effected by the discs and pointer seen at *d*, *d*. After the cut starts, the heads are gradually separated by mechanism that is only partly indicated on the drawings, but which is derived from the rod M lying beneath and parallel with the screw C. This actuates a pinion *c* engaging with a

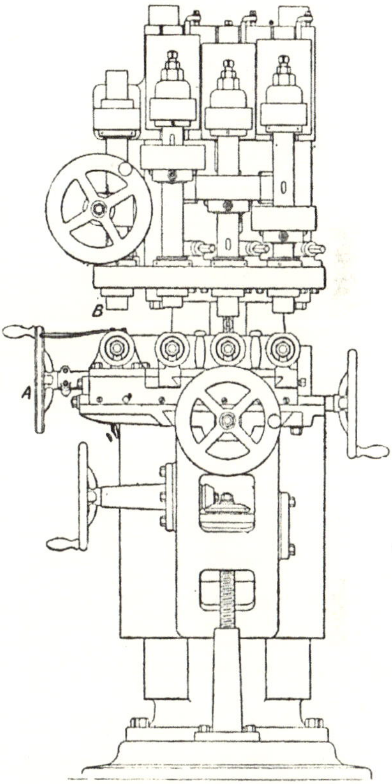

Fig. 126.—Three-Spindle Machine.

rack, whence connection is made by levers with the nuts that operate the heads.

The centrifugal pump N draws its supply of lubricant from the tank O, being driven by the pulley P. The pipe arrange-

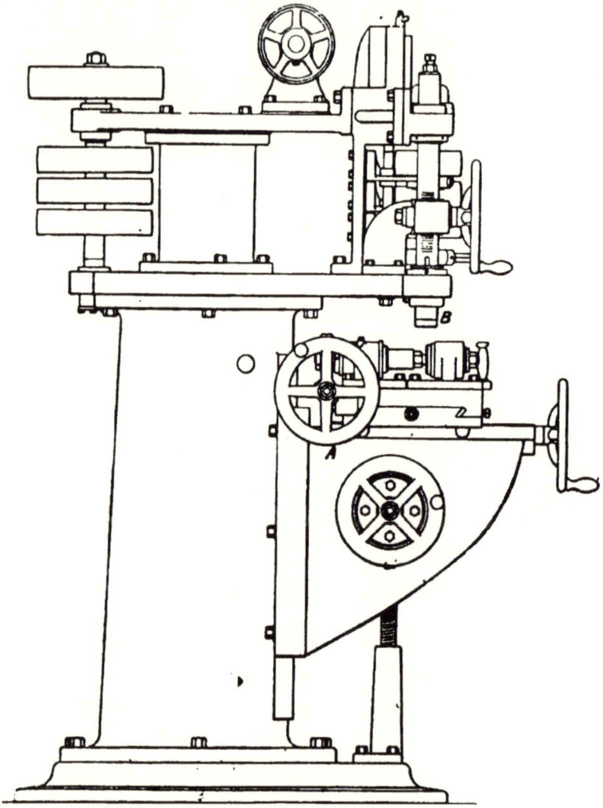

Fig. 127.—Three-Spindle Machine.

ments include a flexible tube connected up to the cutters. The drainage back into the tank takes place through the pipe *g*, but chips and dirt are discharged into the box Q, which, like O, can be pulled out at the front like a drawer to be emptied and cleaned out.

Three-Spindle Machines.—The Kempsmith Manufacturing Company have a fitting to their pillar and knee machines which permits of the simultaneous operation of three vertical milling spindles. The particular application of this device is the milling of the tee slots in machine tables.

The drive is effected by an extra or auxiliary slide carried between the headstock and the vertical brace in front of the knee. This fitting carries a horizontal spindle in line with the main spindle. Three angle wheels, capable of adjustment along it, drive three similar angle wheels on vertical spindles, which have their

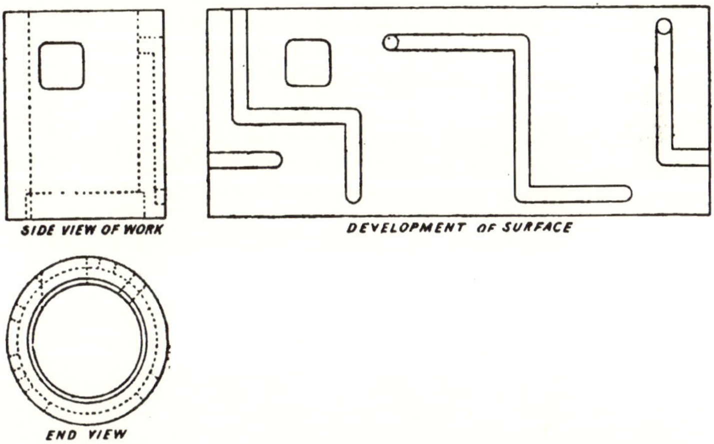

Fig. 128.—Development of Surface of Valve, cut on Three-Spindle Machine.

bearings in slides adjustable along the face of the auxiliary slide. The lower ends of the spindles carry the milling cutters, which, after being adjusted for centres, are driven simultaneously through the angle wheels. The work is bolted to the table and so traversed under the cutter.

Figs. 126 and 127 illustrate a vertical three-spindle machine by the Beaman & Smith Company for duplicating small pieces, and Fig. 128 illustrates the development of the surface of a valve so cut. The periphery of the valve was divided into degrees, and the location of each cutting was fixed by a graduation of the hand wheel A, shown on the fixture in the machine. There is a driving spindle B,

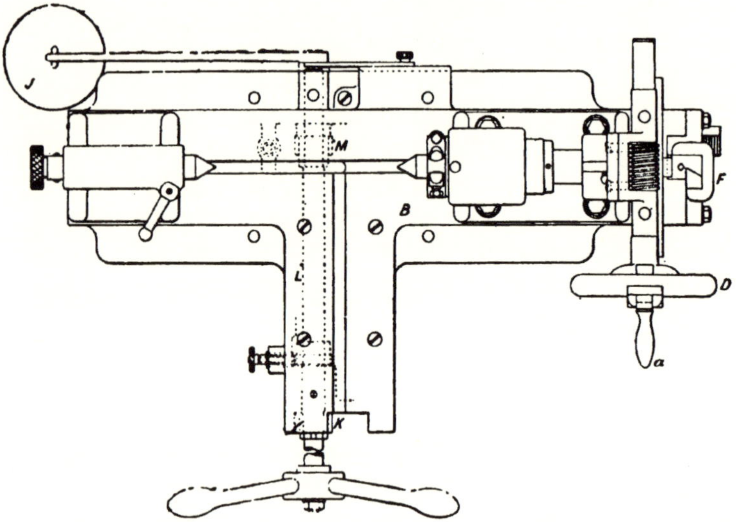

Fig. 129 —Plan View of Cam-cutting Attachment.

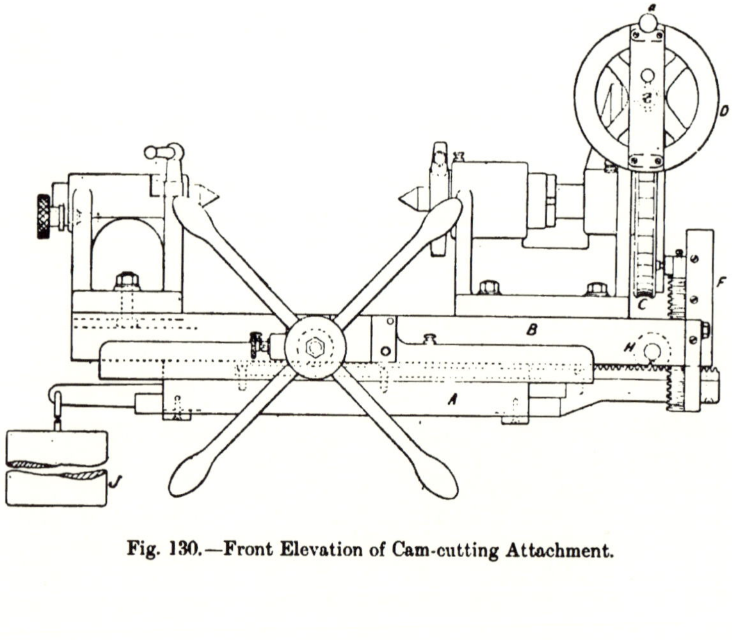

Fig. 130.—Front Elevation of Cam-cutting Attachment.

at the left of the machine which carries a point that runs over the master valve, and on three blanks placed under the operating spindles the cutting is performed simultaneously. This fixture can be removed from the table and any other substituted. The mechanism of the three-spindle machine itself is sufficiently clear from the drawing.

Cam-cutting Attachments.—A cam-cutting attachment by the Brown & Sharpe Manufacturing Company is shown in Figs. 129-131. It is mounted on a base A, which is bolted to the table of the machine. A carries a sliding table B, which is capable of longitudinal travel, and upon which the head and footstock shown are bolted. The spindle in the headstock is driven by the worm and worm wheel C, seen at the right-hand end, the worm wheel being driven by the hand wheel D. The handle *a* of this wheel can be drawn out to give sufficient leverage for turning heavy work round. The outer face of the worm wheel has two tee slots *b*, *b*, at right angles with each other, against which the former is bolted. This is of flat steel, or cast iron, and has its edge shaped to suit the cam to be cut. At the end of the slide B is a vertical guide F, in which slides a rack, having a hardened steel roll at its upper end that comes in contact with the edges of the former. The rack engages with a pinion H, turning on a stud in the slide B, which pinion is also in gear with a horizontal rack, attached to the base A. Hence when the vertical rack is forced down-

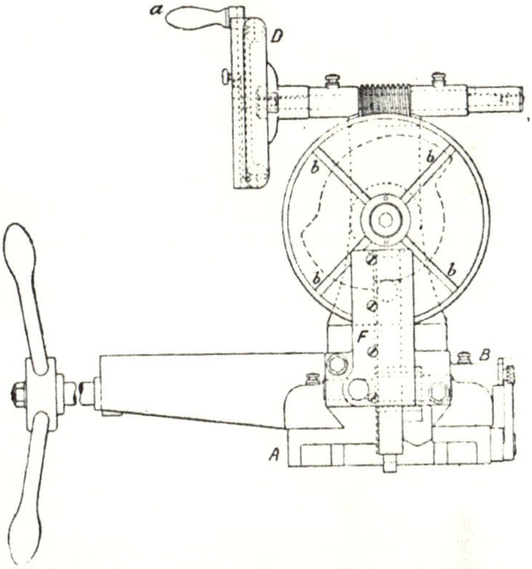

Fig. 131.—End View of Cam-cutting Attachment.

ward, the slide B is moved to the right, and when pressure is removed from the rack, the slide B is pulled to the left by the weight J, and the vertical rack is run up. The roller is therefore always held in contact with the former. And also when the work is rotated by the hand wheel D, the former in turning with the hand wheel produces the longitudinal movement in the slide.

When a cam is being cut on a cylinder, the arrangement is that shown in the figures. But when a face cam is being done, the headstock is set round with its spindle parallel with that of the machine. The vertical rack and guide F being carried round with

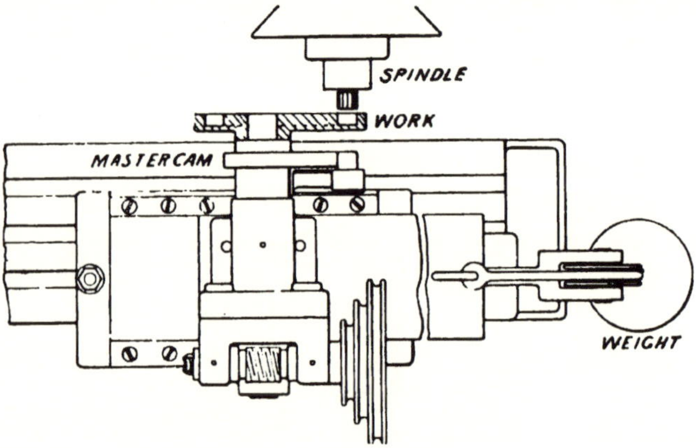

Fig. 132.—Cam-cutting Attachment.

the head, the rack then engages with the pinion K, Fig. 129. This pinion is fast on a shaft L, which carries on its inner end a pinion M, that engages with the horizontal rack in the base A, operating the slide B, as already described, when the headstock is at right angles with the machine spindle.

An attachment for cutting cams on the machine table is shown in Fig. 132, an example by the Cincinnati Milling Machine Company. It is shown at work on a face cam. The arrangement comprises a mandrel to carry the work, and the master or former-cam behind. This mandrel is revolved at a suitable speed, and, by means of the weight seen to the right, the entire slide is pulled constantly sideways, so that the master-cam presses against the roller.

Profiling Mechanism.—The profiling mechanism of a machine by Messrs Webster & Bennett is illustrated in Fig. 133, the piece being a cycle fork crown A. The first Figs., 1 and 2, show it attached to the cam plate B, and just commencing

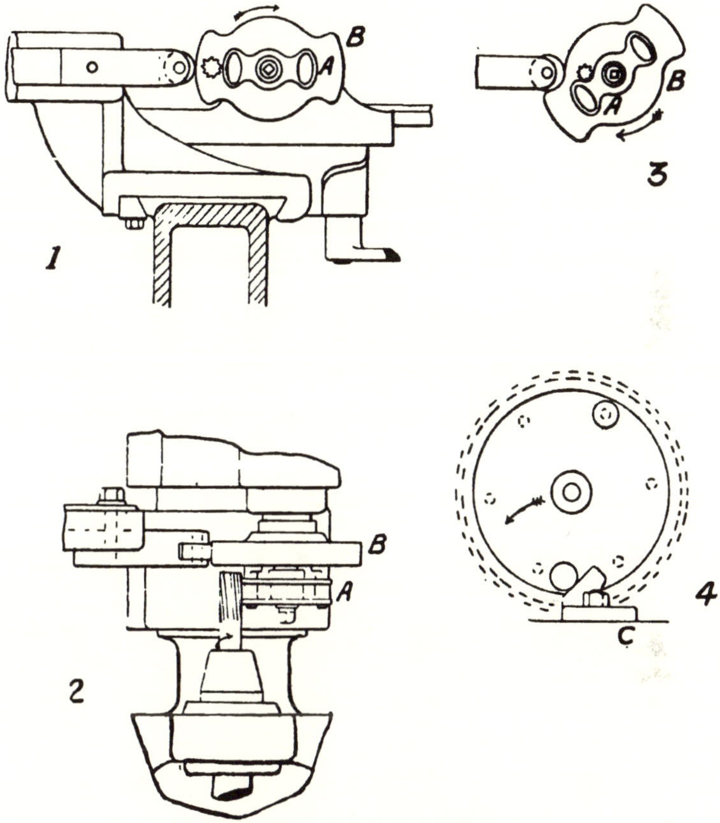

Fig. 133.—Profiling Mechanism.

its revolution in the direction of the arrow. The cam plate B is held against the roller in the ordinary way, until the position shown in the next 3, is reached, at which point the cam is assisted by a roller at the rear of the automatic head, in the manner illustrated in 4. An inclined bracket C is bolted to the bed, against which the roller pushes. The result is that as the bracket will not move, the whole head is pushed over bodily in the direction of the arrow. This has the effect of

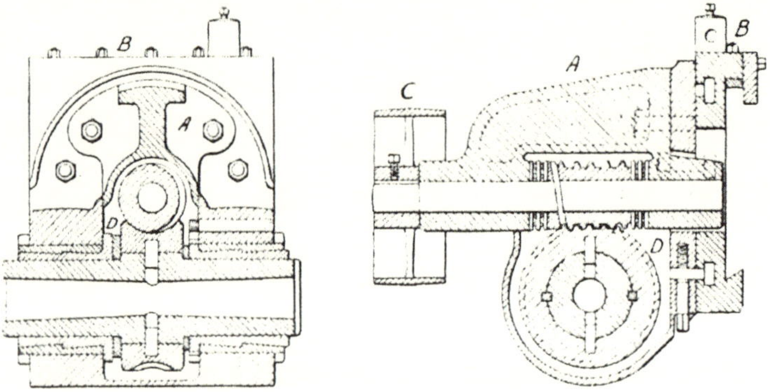

Fig. 134.—Adams' Planer Attachment.

drawing the cam out of the awkward position of the previous figure, and allow it to proceed on its revolution, the large wheel being mounted on a friction cone to permit it to slip slightly in the operation, the lost motion being taken up in a division plate, the small holes in which are indicated in the last diagram 4.

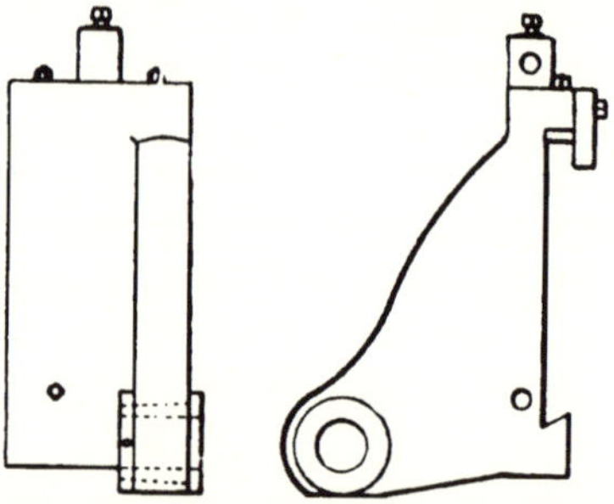

Fig. 135.—Bracket for Tail End of Arbor.

Milling Attachment to Planer. — The illustrations, Figs. 134 and 135, are those of an application of the milling head to the planer, by the Adams Company, the drawings showing the sectional details of the spindle drive.

When in use, the ordinary tool box of the planer is run to the end of the cross slide, out of the way, and the milling head takes its place, being adjustable on the cross rail. The spindle-head A swivels by an annular tee groove on the slide B, so that either vertical, horizontal, or angular milling can be done. There is also a bracket to afford support to the tail end of the arbor in horizontal milling, Fig. 135.

In Fig. 134, C is the driving pulley, which is operated from a separate countershaft, distinct from that of the planer. As the axis of the pulley stands at right angles with that of the cutter spindle, the direction of drive is not affected by the angling of the spindle. The length of the belt is, but the slack is taken up by a weighted pulley. The drive to the spindle is through worm gear D, the worm being of hardened steel, and having ball-bearing thrust collars, and the worm wheel of bronze, enclosed in an oil chamber. The spindle is tapered at both ends, so that cutters can be used at either end. Or, when two heads are used, the cutters can be placed to right and left. The inner bearing sleeves are of bronze. One has a flange at its inner end, and the other a square thread to receive a steel ring at that end, by which end play is taken up. The sleeves have a longitudinal slot, which is fitted with a compressible metal at the outer end, but retains the oil elsewhere. The outer sleeves, of cast iron, are split, and are threaded at the outer ends to receive the ring nuts seen. The outer sleeves fit the bore of the casting, and are fitted to the inner by a double taper. Wear is taken up by turning the ring nuts on the outer end of the larger sleeves. The countershaft has two belts for different speeds. As the table of the planer must be driven at a much slower speed for milling than when planing, the planer countershaft is driven for milling through worm gear, the worm wheel being on this shaft. The worm can be thrown out by a lever, and when out, the table can be returned at the quick planer speed. Feeds are made variable through a friction disc, the roller being on the worm shaft. Feeds up to five inches per minute are thus obtainable.

Machine for Elliptical Holes.—Fig. 136 illustrates a milling machine for cutting elliptical or round holes in boiler plates, made by Curd Nube, of Offenbach-on-Main, Germany.

Cutting the manholes in boilers and boiler plates, and round holes in boilers, is still in many cases effected by hand, with hammer and chisel, or with apparatus in which an analogous process is employed. The machine permits of cutting such holes automatically, by means of a rotary cutter with teeth of sufficient length to enable it when working on plates, or the periphery of cylindrical boilers, to reach through the whole thickness of the material to be cut. The method of operation of the machine is as follows:—

A spindle *b*, carrying a rotary cutter *c*, is driven from an overhead countershaft, through suitable gearing *a*, *a*. The bearings of the cutter spindle *b*, and the intermediate shafting *d d*, are so arranged that the cutter *c* can be moved both vertically and horizontally in any direction. The cutter spindle is carried in an

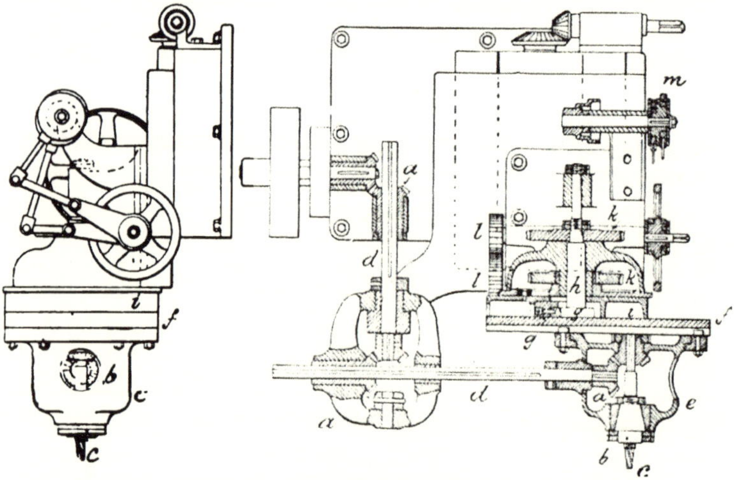

Fig. 136.—Milling Machine for Cutting Elliptical or Round Holes in Boiler Plates.

arrangement *e* in a circular guide, which, with the guide, can be shifted and fixed in position on a plate *f*. This device permits the cutting of circles or manholes of varied sizes. The motion of the cutter *c* is obtained by means of an eccentric *g*, *g* with adjustable throw; moved by a spindle *h*, while a ring *i* to which the plate *f* carrying the bearing *e* of the cutter spindle *b* is fixed, revolves round this eccentric, in the opposite direction. This rotation in opposite directions of the spindle *h* of the eccentric *g*, *g* on the one hand, and the ring *i* with the plate *f* fixed to it, and the cutter *c* on the other hand, is obtained by worm wheels *k*, and worms, and gears *l*, *l*. This arrangement, and hence that of the cutter *c*, is driven from a double-throw eccentric which receives its motion from the driving shaft of the machine, so that when the latter is thrown out of gear this movement is also stopped. The double-throw eccentric *m*, *m* for circular or elliptic motion is itself variable as to stroke, so that small holes can be cut in a shorter time, larger ones in proportion with the longer travel in a longer time.

The machine is arranged for 100 mm. ($3\frac{15}{16}$ inches) difference of diameters, but it may also be adjusted for other differences. Ovals can be cut up to 450 × 350 mm. and circular openings up to 400 mm. diameter. The time within which a normal manhole, 12 in. × 16 in., can be cut in an 18-20 mm. boiler plate is about $1\frac{3}{4}$ to 2 hours; a 10-inch circular hole can be cut in $\frac{3}{4}$ to 1 hour. To do this by hand would take from 6 to 7 and from 4 to 5 hours respectively. All that is required before cutting is to trace a cross representing the major and "minor" axes of the ellipse on the piece of work, and to drill a hole at the end of one of these axes, to give the cutter a starting point. The machine as shown in the figure is fixed to the wall, and the pieces of work go on the ground. The weight of the machine is 1,000 kilogrammes.

CHAPTER VII.

CUTTERS.

Differences in the Teeth of Milling Cutters and Single-edged Tools—Size of Cutters and Spacing of Teeth for Roughing and Finishing Cuts—Rake and Clearance—Spiral Form—" Handing " of Spiral—Compensation for Wear—Attachment to Spindles—Inserted Tooth Cutters—Various Examples—Manufacture of Cutters—Steel—Hardening—Cutting the Teeth—Examples—Grinding and Sharpening—Examples—Clearances—Form Grinding.

Differences in the Teeth of Milling Cutters and Single-edged Tools.—It is not necessary in this chapter to give drawings of all the numerous forms of milling cutters made, even if the scope of this book permitted it. Illustrations of the principal kinds occur in this chapter dealing principally with the tooth forms, and the care of the cutters generally.

In several respects the forms of the teeth differ from those of single-edged tools. They are generally much weaker, approaching in most cases to the form of saw teeth, namely, triangular. They seldom have any front rake, nor ever more than a very slight amount, while most single-edged tools have several degrees of front or top rake. The amount of clearance is generally the minimum of that present in single-edged tools. The spiral or twisted tooth adopted on all edge mills, excepting those of about an inch in width and under, has its analogue in the diagonal or shearing devices embodied in other cutting tools. The mills with staggered teeth, whether solid or inserted, are in effect assemblages of single-edged tools. All milling cutters except the last-named differ from single-edged tools in their fine cutting. They are not in any sense roughing tools, though the term " roughing cuts " is used to distinguish the relative difference between a first and a finishing cut. As they do not and cannot penetrate deeply, there results one of the chief difficulties incidental to these cutters, namely, the inefficiency of a mill which is not ground accurately and supported on a stiff arbor carried in a stiff machine. If the depth of cut is divided between the number of teeth, it is clear that the minute fraction apportioned to each tooth may well exceed the degree of radial inaccuracy resulting from indifferent grinding. How fine this inaccuracy may be is evidenced by the revolution mark visible in all milled work.

The revolution mark is due to the impossibility of grinding teeth exactly alike, because the grinding wheel itself becomes smaller as the work proceeds. Though the difference is extremely minute, it is enough to account for the effects seen.

Size of Cutters and Spacing of Teeth.—The size of milling cutters and the spacing of teeth are two questions that arise first for settlement. With regard to the first, the smaller cutters should generally have preference over the larger, because their traverse is less for a given length of surface cut. The smaller cutter has to move a smaller distance than the larger, and therefore occupies less time in reaching to, and receding from the work.

Messrs Brown & Sharpe state that a difference of half an inch only in diameter has been found to make 10 per cent. in the cost of the work in their practice. In some cases a small cutter is objectionable, as sharpening the edges of milling cutters with a small wheel produces concave edges, but that may be got over by angling the wheel.

The number of teeth in a cutter is now nearly fixed in practice by pitching them at from $\frac{1}{4}$ inch to $\frac{3}{8}$ inch apart. Early cutters were pitched too finely, with the result that crowding and choking occurred. It is well open to question now if coarser pitching than is usual would not be advantageous in roughing cutters and in those for brass, though the practice is to use the same cutters for all metals and alloys, and for roughing and finishing.

Mr Addy's rule for the pitch of cutters from 4 inches to 15 inches diameter is—

$$\text{Pitch in inches} = \sqrt{\text{diameter in inches} \times 8} \times 0{\cdot}0625.$$

In another method, from a German source, the thickness of tooth is first obtained, and thence the number of teeth and pitch, thus:—The diameter, D, of the cutter is first obtained. This depends upon the work to be done, and should be as small as possible. The thickness of teeth, t, should then be determined from the following formula that has been derived from practical experience:—

$$t = \frac{D}{10} + c.$$

For cutters up to	-	-	2 inches diameter		$c = \frac{3}{16}$	inch.
„	from 2 inches to		4 „	„	$c = \frac{5}{32}$	„
„	4	„	$4\frac{3}{4}$ „	„	$c = \frac{1}{8}$	„
„	$4\frac{3}{4}$	„	6 „	„	$c = \frac{1}{16}$	„
„	6	„	$7\frac{1}{4}$ „	„	$c = \frac{1}{32}$	„
„	$7\frac{1}{4}$	„	8 „	„	$c = 0$	„

In order to determine the number of teeth, the nearest whole number to the quotient of $\frac{3{\cdot}1416\,D}{t}$ should be taken. Thus for a cutter of 4 inches diameter we should have 24 teeth, and this would cause a modification of the calculated thickness of the tooth from 0·55625 inch to 0·5027 inch.

Spacing of Teeth for Roughing and Finishing Cuts.— In the spacing of the teeth of milling cutters sufficient allowance is not made for the difference between roughing and finishing cuts. It has been demonstrated that coarsely pitched cutters absorb less power than those of fine pitch. Tests were made in a Cincinnati motor-driven machine on two cutters of the same diameter, but one having twice the number of teeth of the other. The results are given in the table below:—

	Finely Pitched Cutter.	Coarsely Pitched Cutter.
Diameter of cutter	4 inches.	4 inches.
Number of teeth in cutter	30	15
Width of cutter	$\frac{9}{16}$ inch.	$\frac{9}{16}$ inch.
Depth of cut	$\frac{7}{8}$ „	$\frac{7}{8}$ „
Number of teeth in contact with metal	5	3
Volts	110	110
Amperes	13·5	10·5
Feed of table	0·134 inch.	0·134 inch.
Revolutions of cutter per minute	40	40
Thickness of chip per tooth	0·0044 inch.	0·0088 inch.

The conditions, it will be noted, were precisely alike in respect of size, speeds, feeds, &c., only that while one cutter had 30 teeth, the other had 15. But the 15-toothed cutter took a chip 8,000ths of an inch in thickness, while the 30-toothed one took a chip of only half that thickness. But the latter required 13·5 amperes, while the former took but 10·5 amperes, a difference of about 22 per cent. in favour of the coarse-toothed cutter. As the 30-toothed cutter had 5 teeth always in contact with the work, while the other had only 3 teeth in contact, the differences above-named were due evidently to the better clearance afforded to the chips. The case is therefore analogous to the coarse pitching of the inserted tooth cutters and those with staggered teeth.

Wide spacing is favourable to heavy feeding, because the chips escape freely, so that the advantage both of deep cutting and of free cutting is secured. Finishing cuts demand fine-toothed cutters to leave a smooth surface. Hence, though it is not usually done, a difference should be made in cutters for roughing and finishing when the amount to be removed in the first cut or cuts is considerable. For much work, where the total amount to be removed is small, the distinction is of no importance. But when milling comes into rivalry with the planer and shaper for heavy work, the difference ought to be made if the best results are to be secured.

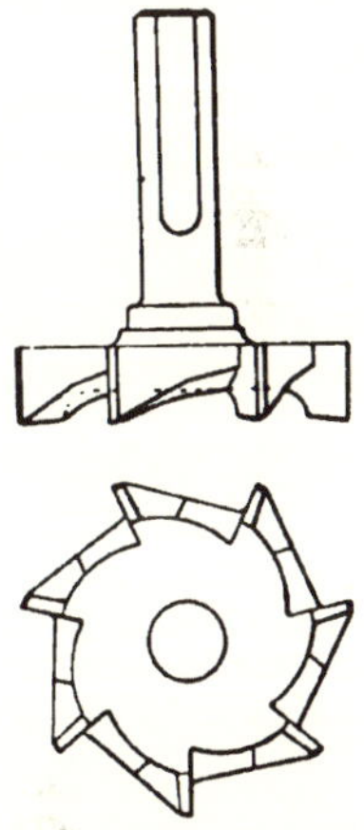

Fig. 137. — Coarsely Pitched Cutter for Gun-metal Fittings.

Fig. 137 shows a cutter designed by Mr James Vose, and used in the shops of Messrs Meldrum Brothers Ltd. for milling gun-metal fittings. It will be recognised as simply a series of brass scraping tools, very coarsely pitched, to leave room for the cuttings to clear. Mr Vose says that more than seven teeth only seems to produce useless rubbing on the finished surface.

Rake and Clearance.—Angle of front rake is not generally imparted to the teeth of cutters. It is found that for ordinary duty radial teeth cut as well as those with rake. Undercut teeth score with slow speeds, and coarse feeds, but not in the opposite conditions. Experiments were made by Mr E. G. Herbert, of Manchester, with three cutters 6⅝ inch diameter, $\frac{9}{32}$ inch thick, with 1-inch hole. In each case there were 30 teeth, which in one cutter were radial, and in the other two were undercut 10° and 15° respectively. The cutters, running at 20 revolutions per minute, or a cutting speed of 42 feet per minute, were fed separately into the vertical faces of heavy blocks, first of cast iron, and then of mild steel, until the machine stopped. The feeds varied from 1·4 inch per minute to 0·16 inch per minute, and there was no perceptible slip of the feed belt.

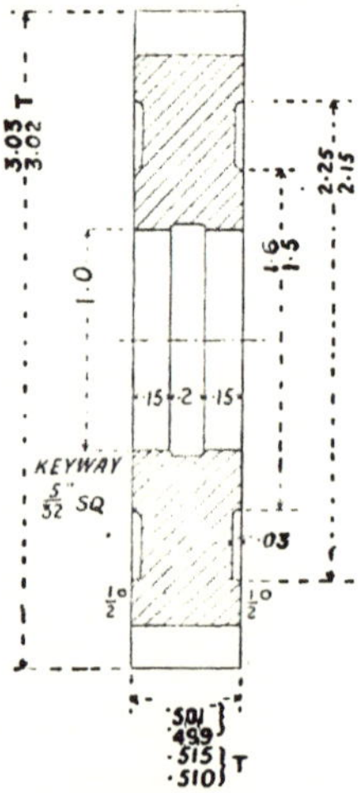

Fig. 138.—Standard Cutter.

The results of the tests were somewhat complicated, but they failed to show any decided advantage of one tooth over another. The general indication seemed to be that the advantage rested with the undercut teeth with a big feed, and the radial teeth gave better results with a fine feed.

A very slight rake is, however, given in some ordinary cutters, ranging from 2° to 5°, the latter for wrought iron. This, however, should not be adopted in formed cutters, which have to be ground on their faces.

The angle of clearance (which is that given by grinding the top edges of the cutters) ranges from 3° to 8°, the latter being rather excessive unless used habitually on brass. This amount is fixed by the setting in the grinding machine by the tooth rest.

The two broad groups into which milling cutters are divided are the axial, in which the teeth are in planes parallel with the axis; and the end, or radial, in which they are at right angles with the axis. There are many sub-types, as for instance tapered mills and profiled mills in the first, and convex-ended and other forms under the second.

The axial mills generally go on arbors. Fig. 138 is from a shop drawing giving a section of a cutter of this kind, with Mr Brayshaw's remarks thereon below.

"According to this drawing, it has been determined that if the error in the thickness of a ½-inch cutter does not exceed one-thousandth (0·001) of an inch, it is good enough. This is clearly shown, and the grinder must adhere to the limits given, but must not waste time in making every ½-inch cutter to within one half-thousandth (0·0005) of the nominal size.

"Again, it has been found that about one-hundredth (0·01) is a reasonable allowance for cleaning out the turning marks on the sides after hardening. It is, however, quicker to grind off a few extra thousandths than to turn them off, and the turner must keep within the limits—ten to fifteen thousandths above ½ inch $\left(\frac{0{\cdot}515}{0{\cdot}510}\right)$. He has no excuse for leaving too much or too little for grinding, nor yet for wasting time by taking a cut of two-thousandths (0·002) off the side.

"It is shown that the actual diameter is not important, and the turner has a limit of one-hundredth (0·01) of an inch, which he must keep within. No grinding size is given here, which means that the grinder must just clean out the turning marks.

"The drawing shows that the side recesses may vary in diameter by one-tenth (0·1) of an inch. The clearance each side is stated as ½°, and it is essential that this shall run out to the extreme tips of the teeth."

Messrs Brown & Sharpe give about five one-hundredths in one inch for clearance for mills that have to cut grooves. That is, a grooving mill should be about one-hundredth of an inch thinner at one inch from its circumference than it is at the edge. The firm gives a limit of two-thousandths in thickness to their grooving mills. Mr Brayshaw's ½° on each side causes the cutter to become two-thousandths (0·002) thinner when ¼ inch has been ground off its diameter. More clearance would cause it to lose its proper thickness too soon. Extra clearance might be given to cutters when the width of grooves is not particular, or when they are used only for roughing.

Spiral Form.—The spiral form given to all but the narrowest axial mills varies from about 15° to 40°. Its object is to reduce the stress on a tooth by causing it to cut in detail, in a shearing fashion. Its value is more pronounced in the fibrous metals, as wrought iron and mild steel, than in cast. In ordinary axial mills it may be low or high in amount, and right or left handed in direction. But in end mills, and those which cut profiles, conditions come in which fix these things arbitrarily.

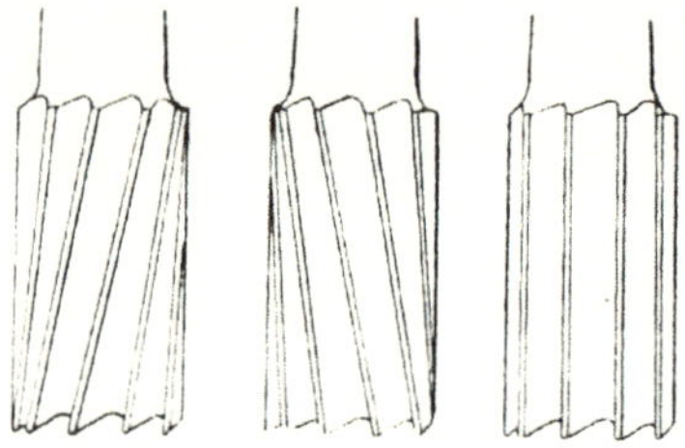

Fig. 139. Fig. 140. Fig. 141.
Examples of End Mills.

If cutters operate by their sides only, the inclination may be large, ranging up to 30° or 40°, which in wrought iron gives a marked advantage. But in end mills this should not exceed 20°, because the teeth would be too keen, and become broken, due to the large angle of front rake imparted to the end teeth.

"Handing" of Spiral.—The question of which hand the flutes in a milling cutter should run may seem of little moment. In the case of end mills going deeply into the work it seems best that the flutes of the sides should be right handed, like those of twist drills. These then tend to draw the chips out of the groove, while left-hand spirals would tend to force them down towards the cut.

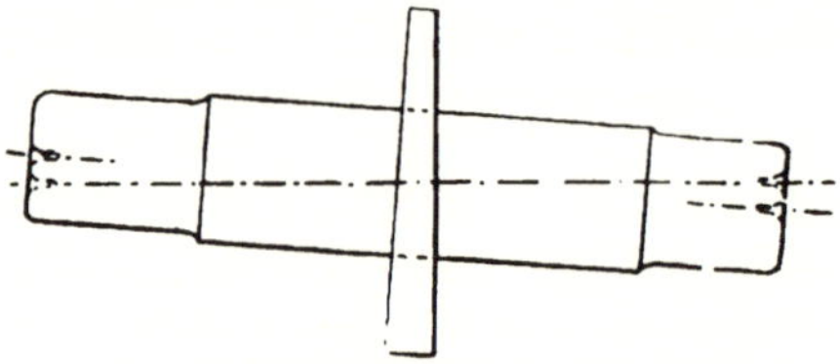

Fig. 142.—Mandrel Set for Turning Diagonal Joints of Cutters.

The handing of the twist is important in certain cases. Fig. 139 is a right-hand spiral with right-hand teeth, Fig. 140 is a left-hand spiral with right-hand teeth. The differences in the two in working are these:—Fig. 140 is only suitable for edge milling, since the end teeth have negative rake, and the direction of twist tends to push the cutter into its socket, and the work down to the table, and to thrust the cuttings downwards. But for use as an end mill, the form in Fig. 139 is correct, because the teeth have front rake, and the tendency is to draw the cuttings upwards. For general end milling, the cutter in Fig. 141 is more often used, giving no trouble of any kind.

Compensation for Wear.—When it is desired to make compensation for wear in side mills, they are jointed diagonally, or the

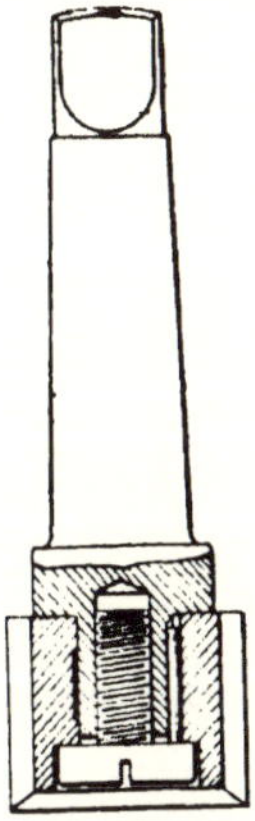

Fig. 143.—Common Method of holding Shell Cutter on Arbor, with Round Key to Drive, and Screw to Hold on.

Fig. 144.—Shell Cutter Arbor, with Nut for Forcing off the Cutter.

teeth are interlocked, or sometimes every alternate tooth on the meeting faces is cut away to make room for the teeth on the other mill. Either of these permits of the insertion of thickness slips of

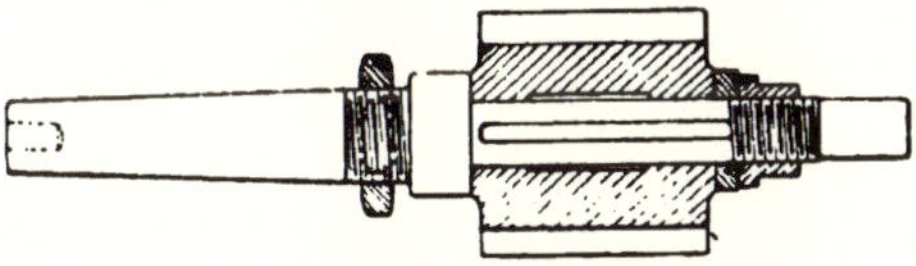

Fig. 145.—Cutter held on Arbor with Nut, Arbor forced out of its Hole with a Nut.

paper or other material. Diagonal jointing of smooth faces is not open to the objection of leaving a mark on the work, which parallel flat cutters would do, if jointed. The diagonal jointing is done by turning the faces on a mandrel, Fig. 142, having two sets of centres, one for turning the parallel faces of the cutters, the other situated at about $\frac{3}{16}$ inch out of centre, used for turning the diagonal faces. When the faces are tooled, the two portions are put together and united with a pin, the key groove is cut, and the teeth milled, hardened, and ground.

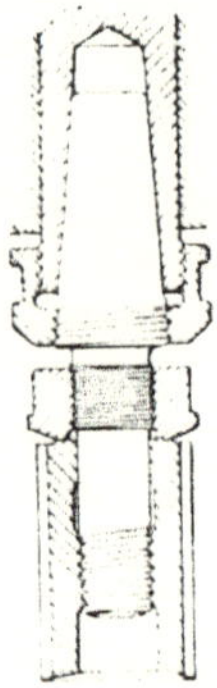

Fig. 146.—Muir's Coupling Mandrel.

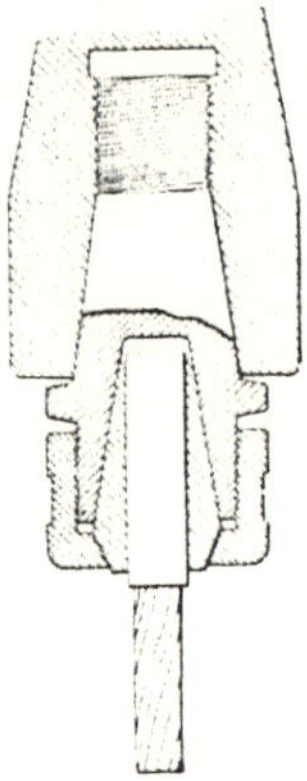

Fig. 147.—French Method of Gripping Cutter with Parallel Shank.

Attachments to Spindles. — Mills fit into taper sockets in their spindles, or they fit over arbors. Small mills are used in both ways, but large ones, and also the greater number of any size employed, fit over arbors only.

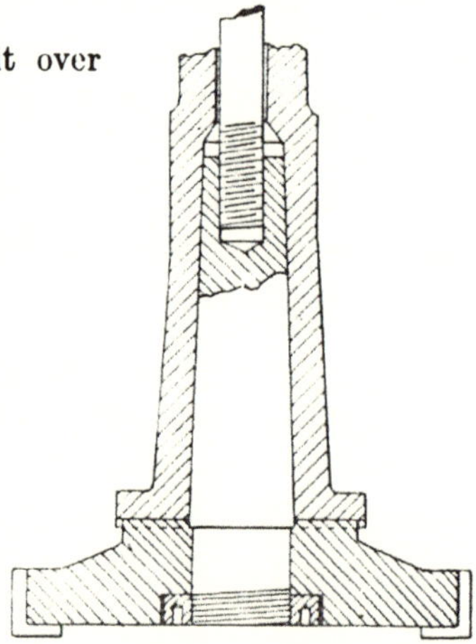

Fig. 148.—Large Face Cutter locked with Clutch and Ring Nut.

Figs. 143-148 illustrate taper sockets for mills, which are nearly self-explanatory. Solid shank mills are seen in Figs. 139-141. Fig. 143 shows the common method of holding shell cutters on an arbor, using a round key to drive, and a screw, the head of which lies in a recess in the mill to secure it. In the next example, Fig. 144, a nut is fitted above instead of a plain collar. The object of this is to push off the cutter by the simple pressure of turning the nut, a feature appreciated if cutters stick. In Fig. 145 a nut and washer are shown, with a forcing-off nut. Fig. 146 is Muir's coupling.

The arbor is retained in place with a screw collar, the cutter is threaded to receive the arbor, and a lock nut above holds it securely. A method of holding a parallel shank cutter in a split tapered grip with an encircling nut is seen in Fig. 147. The arbor is retained by a screw on its end. In Fig. 148 a cutter of large diameter is driven from the spindle end by means of an interlocking clutch, and held with a ring nut turned with a tommy.

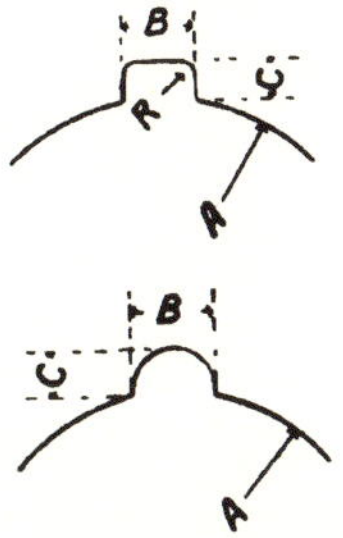

Fig. 149.—Diagrams for Proportions of Key Ways.

Very thin mills are ground straight through to fit the arbors, and wider ones are recessed as in Figs. 138 and 145. The key groove should be of half-round section, to lessen risk of cracking in hardening. Some, however, have square grooves with radii. The thinnest cutters are not keyed, but pinched between shoulders. The diagrams, Fig. 149, and tables below give standard sizes of both shapes of key ways.

SQUARE KEY WAY.

Size hole, A	$\frac{3}{8}$ to $\frac{9}{16}$	$\frac{5}{8}$ to $\frac{7}{8}$	$\frac{15}{16}$ to $1\frac{1}{8}$	$1\frac{3}{16}$ to $1\frac{3}{8}$	$1\frac{7}{16}$ to $1\frac{3}{4}$	$1\frac{13}{16}$ to 2	$2\frac{1}{16}$ to $2\frac{1}{2}$	$2\frac{9}{16}$ to 3
Width, B -	$\frac{3}{32}$	$\frac{1}{8}$	$\frac{5}{32}$	$\frac{3}{16}$	$\frac{1}{4}$	$\frac{5}{16}$	$\frac{3}{8}$	$\frac{7}{16}$
Depth, C -	$\frac{3}{64}$	$\frac{1}{16}$	$\frac{5}{64}$	$\frac{3}{32}$	$\frac{1}{8}$	$\frac{5}{32}$	$\frac{3}{16}$	$\frac{3}{16}$
Radius, R -	·020	·030	·035	·040	·050	·060	·060	·060

HALF-ROUND KEY WAY.

Size hole, A	$\frac{3}{8}$ to $\frac{5}{8}$	$\frac{11}{16}$ to $\frac{13}{16}$	$\frac{7}{8}$ to $1\frac{3}{16}$	$1\frac{1}{4}$ to $1\frac{7}{16}$	$1\frac{1}{2}$ to 2	$2\frac{1}{16}$ to $2\frac{7}{16}$	$2\frac{1}{2}$ to 3
Width, B -	$\frac{1}{8}$	$\frac{3}{16}$	$\frac{1}{4}$	$\frac{5}{16}$	$\frac{3}{8}$	$\frac{7}{16}$	$\frac{1}{2}$
Depth, C -	$\frac{1}{16}$	$\frac{3}{32}$	$\frac{1}{8}$	$\frac{5}{32}$	$\frac{3}{16}$	$\frac{7}{32}$	$\frac{1}{4}$

Inserted Tooth Cutters.—The growing employment of the inserted toothed cutters for face mills has done as much for the extension of the practice of milling as perhaps any other single improvement or innovation. The terms face milling and rotary planing are both applied to this practice. The latter is the more suitable term, because it distinguishes between the single cutters inserted in a disc, and the end milling as done with teeth on the ends of solid cutters.

The ordinary solid mill with finely pitched teeth is never a roughing tool in the sense of removing material in quantity, its depth of cut is far too limited. It roughs, of course, but always in a depth measured in hundredths of an inch. But the coarsely pitched inserted cutters are narrow planer tools with penetrative capacity, and they therefore rough deeply when backed up by machines of sufficient power. They are made and used in all dimensions from a few inches up to several feet, and on vertical and horizontal spindles. There would appear to be no limitations in reason to the size of a face mill. It fills a most valuable place in the work of roughing down, and in that rough finish which is commercially good enough, in perhaps half the volume of engineers' work done. Its utilities lie in the facing of flanges and feet, and of the areas to which they are to be attached. It is of especial value in plates and constructional work, in facing the ends of beams and stanchions, of columns, and pipes, and the edges of plates superimposed, involving numbers of similar pieces. The front rake which is imparted to the cutters in most of these heads makes them true cutting tools, and their coarse pitching permits the chips to get away freely, which is further facilitated by the horizontal setting of the majority of the spindles which carry the rotary planers.

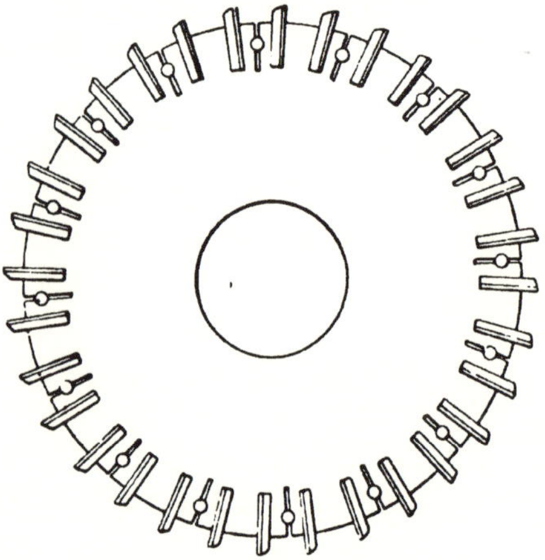

Fig. 150.—Inserted Tooth Mill.

Fig. 150 shows a common and good form of inserted tooth cutter. The body is split, and the cutters are fixed by taper pins driven into every alternate spacing. In various modifications this device of tapered tightening pins is adopted.

A widely used method of holding the blades is shown in Fig. 151. The blade A is ground on the sides, on a magnetic chuck. The bush B is turned parallel, and has a flat milled on it at an angle with the centre line. This bush, which fits in a recess, as shown, is simply a wedge, and is knocked in. There is a screw C to prevent it becoming loose. A second screw D, the patent of Mr W. S. Baskerville, is shown for adjusting the blades sideways.

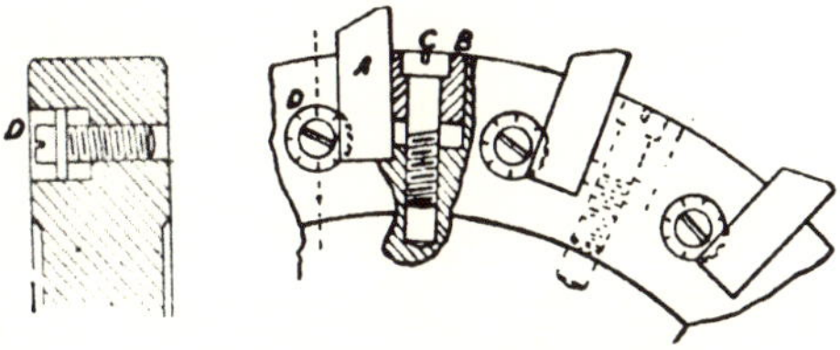

Fig. 151.—Inserted Tooth Mill.

Fig. 152 shows the construction of a large inserted tooth cutter, nine inches diameter, by twelve inches long. The cutters are inserted in diagonal grooves, and secured with set screws pressing a wedge piece against their faces. The cutting faces are radial, and of helical shape. This was produced after the fixing of the cutters in the diagonal milled grooves, the machine being geared to cut the spiral faces.

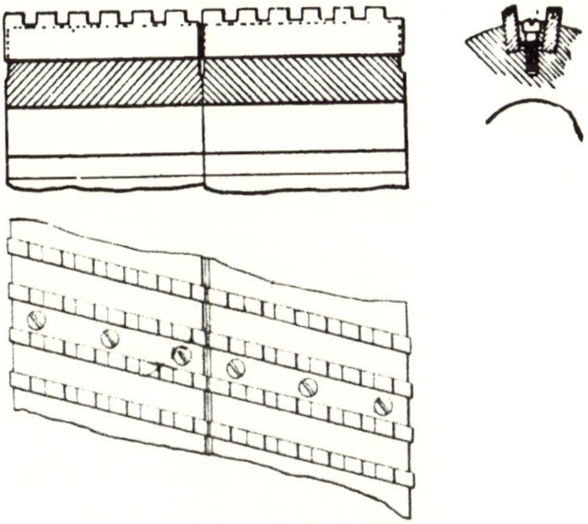
Fig. 152.—Inserted Tooth Cutter.

The tooth gaps are cut to alternate, and not in a spiral fashion. Each alternate cutter was put in place, and the gaps cut at intervals. Afterwards these cutters were removed, the other alternate set inserted, and the grooves cut to come behind the teeth in the first set. The objection to cutting them spirally, as though screw-cut, is that one side of each gap would be prevented from clearing.

Fig. 153 illustrates a novel form of inserted tooth cutter made by the Garrard Manufacturing Company Ltd., of Birmingham. Its main feature is the adjustability of the cutters in the head, to permit of backing off by simple turning. The teeth A have their inner ends convex to fit down in concavities in the head. The latter consists of three pieces, the body or stock B, and the caps C, C, and the latter are clamped against the teeth by the nuts on the mandrel that secure the head. Conical lugs *a*, *a* are provided on the sides of the teeth which enter into annular recesses in C, C, tapered on the outer edges to tighten the teeth by the screwing up of the clamping nuts. When the caps C, C are slackened, the cutters can be turned through a small arc in either direction, to the extent of the difference in their thickness, and the width of their grooves. The cutters are pushed over to one side, and clamped to be turned. They are then removed, and hardened. When replaced they are moved over to the opposite side, which gives the angle required for cutting, when they are again clamped.

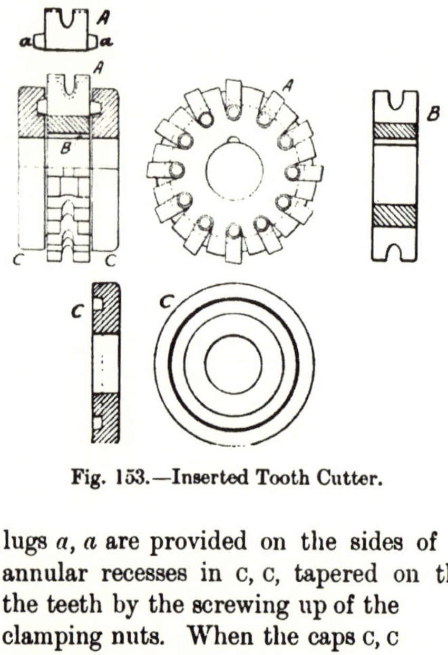

Fig. 153.—Inserted Tooth Cutter.

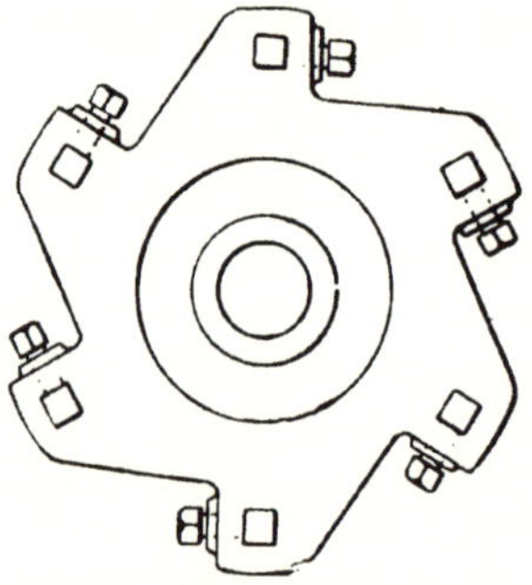
Fig. 154.—Inserted Tooth Cutter.

Fig. 154 is an inserted tooth cutter of French design, in which set screws are used to hold the tool points.

Fig. 155 illustrates a German type of large cutter by the Maschinen Fabrik Lorenz. Provision is made for adjustment of the cutters by means of tapered operating screws (one of which is shown in section adjacent), and clamping is by the outer screwed ring.

Fig. 156 shows an inserted tooth cutter having front rake, used in one of the A. Herbert machines, with swivel head attachment. The fastening of the cutter is by means of splits, similar to Fig. 150.

Fig. 157 shows details of the fitting of a large cutter by the Tangye Tool & Electric Company Limited. The diameter over the

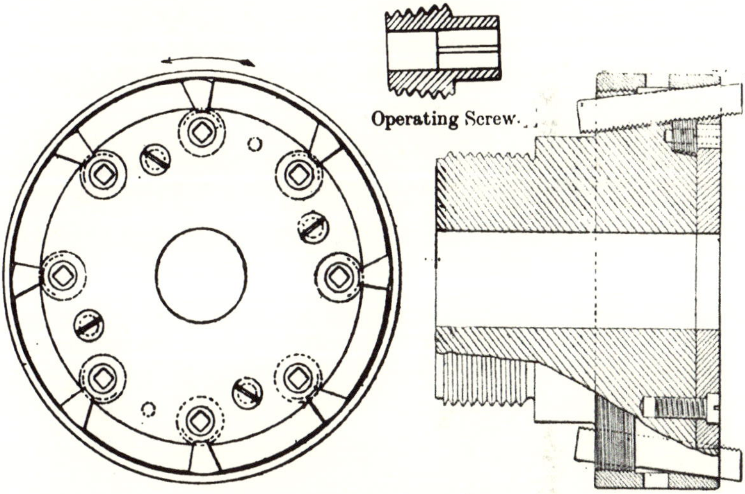

Fig. 155.—Inserted Tooth Cutter.

cutter is 2 ft. 6 in. The block carries thirty-four roughing tools, 1 inch square in section, and two finishing tools, 1½ inch by ¾ inch section. The tools are inserted at an angle of 5° right hand, and are adjusted by means of ¾-inch screws, turned with a podger, inserted in a square hole, and locked with a nut. Two set screws hold each cutter firmly. A pinion rotates the cutter block through an internal ring of teeth. It is used on the style of machine shown on page 137.

The cutter of the Goliath milling machine at the Fairfield Works measures 16 feet from point to point of tool. It has fifty-

Fig. 156.—An Inserted Tooth Cutter on Herbert's Machine with Swivel Attachment.

five roughing and five broad-faced tools. It makes one revolution in 3 minutes 25 seconds. The feed is $2\frac{1}{4}$ inch per revolution, equal to about ·66 inch per minute. The depth of cut is $\frac{3}{4}$ inch. The cut per tooth is 0·04 inch per revolution, omitting the broad-faced tools. This multiplied by the depth of cut makes the cross section of cut 0·03 square inch per tooth. This, with a cutting speed of about 22 feet per minute, shows a good record by comparison with a single planing machine tool.

In some shops where a good deal of face milling is done, inserted cutters are cast in their heads. The experience of people who have tried these differs, some saying that the teeth work loose, and that the teeth cannot be tempered. Others use them

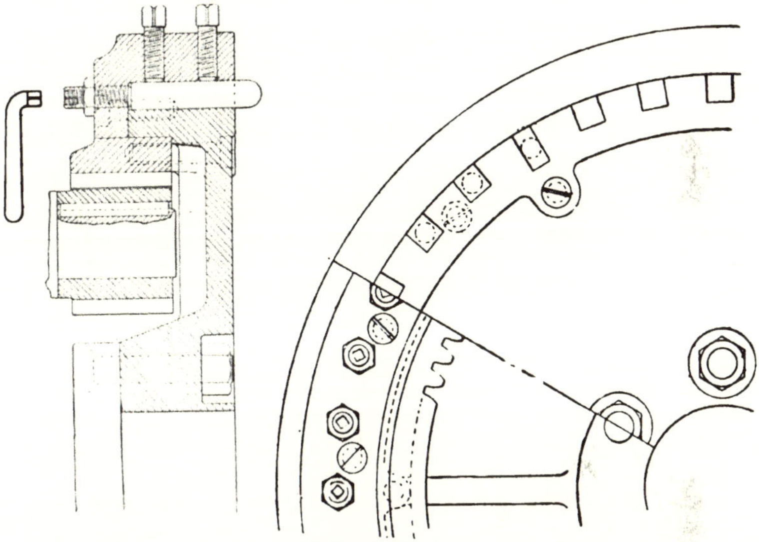

Fig. 157.—Block with Inserted Cutters. (Tangye Tool & Electric Company Limited.)

regularly, and find them satisfactory for finishing as well as roughing. The method adopted is this:—

Either Mushet self-hardening steel is used for the cutters, or ordinary good tool steel. They are cut off to 2 or 2½ inches long, and set with the cutting ends downwards in the bottom of a core box, and rammed round with core sand, which is dried. The core containing the cutters is then dropped into a print impression obtained from the pattern head, and so cast, some extra hot iron being poured through the mould to ensure union, imitating the practice of burning-on. The act of casting anneals the steel, so that the teeth can be trued up after the shank of the head has been turned. Or they can be roughly ground, after which they are finished in the tool grinder in the ordinary way. They are then hardened. In some of these cutters made by Professor Sweet, 10° of face angle or rake is given in setting the cutters in the mould. It is necessary to prevent rust forming in the steel before casting. They are therefore ground bright on the emery wheel, and put directly into the mould; or preferably coated with boiled oil, or else are tinned, either of which methods effectually prevents the formation of rust. The presence of rust, even in traces, causes a blow, and prevents union of the metal.

Manufacture of Cutters.—The question of making *versus* buying milling cutters is often a vexing one. Cutters are not cheap, neither are the lathes, milling machines, and grinding machines by which their teeth are formed. Cutters purchased from an experienced firm are reliable. Those made by the shop tool smith and in the tool room are not always so, either as regards truth or temper. The case is much akin to that of twist drills and reamers: few small firms make their own, but prefer to buy them. Still, the making of milling cutters is work that a firm can get into gradually. Having a universal milling machine in the shop, the teeth can be cut spirally or straight; hardening and tempering should be managed in the tool room, and grinding machines are semi-automatic. Many large firms make their own cutters. The more unusual the forms required, the greater the reason why they should be made on the premises. This is the case in all formed cutters, the outlines of which are almost infinite, and large numbers of which are used in some shops. In such cases it is better to be independent of outside help, which often means delay and misunderstanding of precise requirements.

Steel.—The production of a suitable class of steel for milling cutters is of cardinal importance. It is useless to spend a lot of time on mills, and then fail through buying cheap steel unfit to stand the work demanded of it. The following tables illustrate the chemical composition of high-class steels suitable for mills:—

Crucible Cast Steel.

	Per cent.
Carbon -	1·2
Silicon - -	0·112
Phosphorus -	0·018
Manganese	0·36
Sulphur - -	0·02
Iron, by difference	98·29

Iron -	about 98		
Carbon - -	from	1·0	to 1·5
Manganese	„	0·10	„ 0·40
Silicon	„	0·10	„ 0·25
Sulphur	„	0·003	„ 0·004
Phosphorus -	„	0·01	„ 0·02

Actually when steel is wanted for mills, the manufacturer must be informed of the specific purpose for which it is required, and a good price paid.

The quality of the steel cannot be judged by inspection, nor can cutters made of good or bad steel be distinguished one from the other after manufacture. The temptation to use cheap steel is due, not only to the lessened first cost, but to the lower expense of machining it. The economy of using such steel is considerable on large cutters, sometimes effecting a saving of one-third or one-half; but it should not be allowed to weigh in small ones, particularly those which have profiled edges or which are to be built up and thus involve extra work in the jointing. This is one of the advantages of the inserted types—the ability to use cutters of the best material in the commoner matrix.

After cutters have been shaped by turning, grinding, and milling they should be allowed to rest for a few days previous to hardening. If they are to be annealed, Mr Brayshaw recommends keeping an excess of charcoal near them in the furnace to maintain a reducing atmosphere and prevent risk of their becoming decarbonised.

Hardening.—With regard to the methods of heating and hardening much has been said and written, with much difference in expression of opinion. Two or three essentials are clear—that heating must be gradual and uniform, that overheating must be avoided, that every different brand of steel needs a different temperature, that there is a best and an exact temperature for quenching a steel, that cutters of different shapes require differences in the ways of heating and quenching, and that the virtues of particular quenching liquids are of a somewhat indeterminable character. Some of these matters call for extended remarks.

A gradual heating up is one of the virtues in the process. By it risk of cracking is greatly lessened. Even though heating may be uniform throughout, yet it is better if brought slowly to that condition than rapidly. The importance of these facts is so great that it is desirable to have some precise and mechanical means of regulation instead of the old method of the forge. As the dimensions of cutters have increased, the necessity for the substitution of the mechanical for the rule of thumb has become more apparent. Hence the efforts made to utilise a liquid for heating instead of a furnace heated by air, and employing a pyrometer for registering the exact heat of the liquid or of the heating furnace. Though molten lead has been employed, there are several objections to its use. Whatever means of heating be adopted, the pyrometer is becoming a most valuable instrument for ensuring uniformity of temperature.

Overheating may be present without cracking resulting if the heat be very uniform. It is an axiom that the lowest heat which will give sufficient hardness is the best to adopt. Cracks may be detected by sand-blasting a cutter after hardening, when they will be revealed.

The question of frequency of grinding a mill is one of economical importance. The time lost is that due to changing mills in the machine with stoppage of the latter, the time occupied in grinding, and the more rapid wear and shorter life of a soft mill over that of a suitably hardened one. This is one way to look at the subject of hardening suitably done or improperly.

This, however, may be less serious than the question of hardening cutters with the least risk of cracking. It is less risky to make cutters medium hard than to make them sufficiently hard to be durable, combined with sufficient toughness to avoid risk of cracking.

Cracks are the great evils to which cutters are liable, and the larger a cutter and the more abrupt the forms of its teeth the greater is this risk. Slight distortion does not matter, because that is corrected in grinding; but cracks are fatal, and these sometimes do not develop until the first grinding is done and the skin being removed.

Risk of cracking is lessened by avoiding keen angles in the roots of the cutters. Hence, though the teeth sections may be angular, the roots should have radii. But the important work lies in the methods of heating and cooling adopted. The chief points to be observed after a suitable temperature are equal heating and the avoidance of draughts.

The temperature must be graded to suit a given brand of steel. It may be estimated by the eye or by the pyrometer, but it should not be allowed to fluctuate. Hence the open fire is not suitable for mills as it is for single-edged tools. If employed at all, the cutters must be thrust into a tube; but a gas furnace, such as those used for case hardening, or one of the special furnaces made, is the proper thing. The work is placed in a muffle or in iron boxes, either open or enclosed in hardening mixtures. Some hardeners prefer mixtures for the larger mills, such as charred leather and charcoal pounded to the size of a pea, within which the mill is completely enclosed and kept for three or four hours at a low red heat. The mill is then quenched in a bath of raw linseed oil, being moved about rapidly to bring fresh oil into contact with the teeth, and there kept till cold. Mills with large teeth treated thus need not have the temper drawn subsequently. Smaller ones may be drawn to a light straw or at a temperature of about 430° Fahr. In this packing method, which Mr Markham adopts in preference to open heating and water quenching, a great deal depends on the quality of the leather used. The scrap shoe sole leather from factories is recommended. This should be prepared by packing in one of the hardening boxes, luted with fireclay, and put in the furnace just long enough to allow it to become charred sufficiently to be pounded up well.

Another important point in the work of hardening is to remove internal strains by slightly warming the cutter immediately after quenching. Mr Markham attaches great importance to this in all mills of over ½ inch diameter. The suitable temperature is that at which the hand cannot be held on the cutter, but not so hot as to show any temper colours. The explanation is that which applies to all annealing—the heating of the external hardened parts sufficiently to allow them to adjust themselves to the changes of the interior, which go on for some time after the outside has set rigidly. This reheating can be done either over the fire, or, in the case of the smaller cutters, in a tray immersed in a tank supplied with boiling water.

The difficulty of hardening without cracking is due to mass, and whatever counteracts the effect of this lessens risk. In a large mill the body continues to shrink after the teeth have set, and so tears itself away from some of them, or else the metal remains in a state of internal tension ready to develop cracks on the slightest occasion. Any suitable method of annealing therefore lessens this risk, whether done before or subsequently to hardening. For the same reason all cold draughts and unequal and sudden changes in temperature are to be avoided in the work of heating mills.

The methods of quenching are generally familiar. It is well understood that long cutters must be plunged vertically and flat ones edgewise, done in both cases to avoid warping. The object is to cool all parts simultaneously, and to allow of free access of the water to all parts alike. Similar methods of letting down are adopted as in other tools. As soon as the teeth are chilled they may be lifted out, and the internal heat allowed to diffuse into the teeth and then quenched again, the final quenching being done at a pale straw or medium orange.

Water that has been used repeatedly is considered by old smiths a better medium than fresh water. The reason seems to be that the air is driven off, and that the presence of air prevents free access of the water. Others use warm water raised to from 90° to 100° F. A safe, suitable bath is made by dissolving soft soap in boiling water to the consistency of cream.

Oil is used also. It does not cool so quickly as water. The cutter teeth should be as hard as the teeth of a file.

Milling cutters with holes are best stored on vertical boards studded with wooden pins. These are frequently hinged, with pins on opposite sides, either enclosed in cupboard fashion or ranged down passages along which the man in charge can walk.

Cutting the Teeth.—There is often much similarity between the methods of cutting the teeth of mills and of grinding them. That is, the sectional shapes of the cutters and the grinding wheels are often identical. This only holds good, however, when a tooth space is ground entirely, and not when the tips or the faces alone are done. The following are illustrations of both kinds.

Fig. 158 is an example of a cutter with straight teeth, having its teeth formed with a cutter of the general shape seen in Fig.

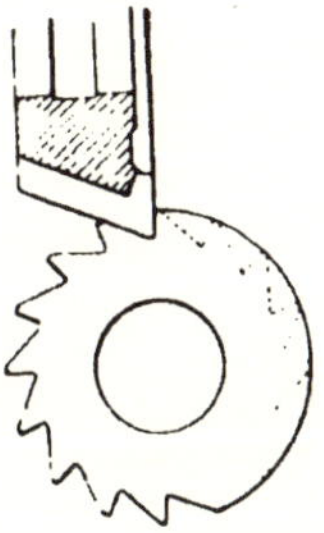

Fig. 158.—Cutting Teeth of Milling Cutter having Straight Teeth.

Fig. 159.—Usual Type of Cutter for Producing Teeth.

Fig. 160.—Deepening Teeth, or Cutting New Teeth with Emery Wheel.

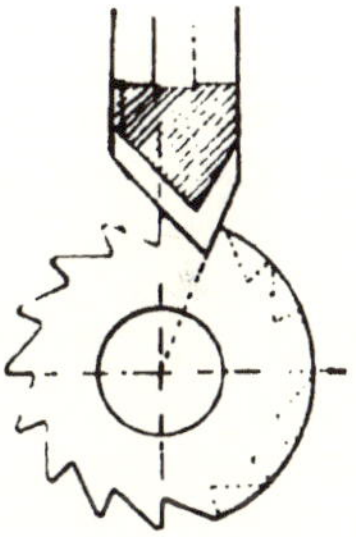

Fig. 161.—Cutter for Spiral Mills.

159. Such a cutter would be sharpened afterwards on its edges, but as it becomes worn its teeth would require to be deepened occasionally with a grinding wheel like Fig. 160, or the teeth might be cut in the first place with an emery wheel, though that is an unusual and unsatisfactory method. These are ordinary cutters, which, if profiled in section, lose their shape by re-grinding. Fig. 161 shows the cutter suitable for a spiral mill with the same form of tooth.

The most valuable improvement introduced into milling-cutter practice, ranking only second to that of the development of special cutter-grinding machines, was that of making "formed" cutters. This term was given by the firm whence they emanated—the Brown & Sharpe Manufacturing Company. The advantage is that the cutters never lose their original sections by regrinding. The grinding is done only on the cutting faces. The profile forms are

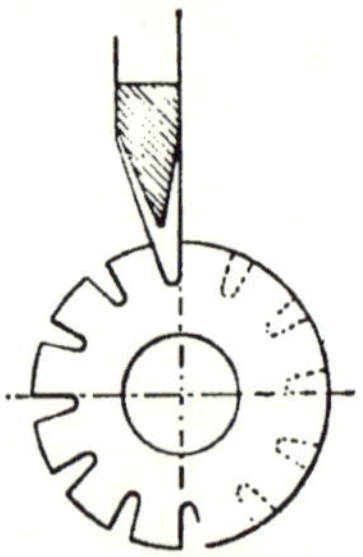

Fig. 162.—Cutting Backed Off Mill having Straight Teeth.

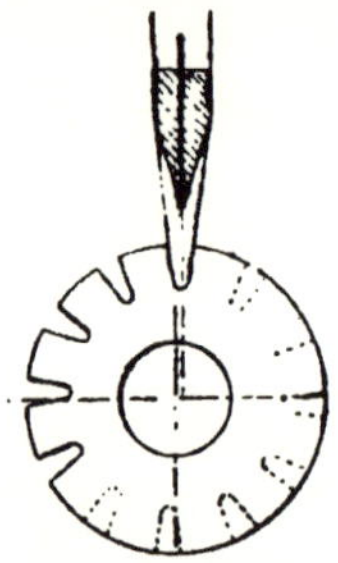

Fig. 163.—Cutting Backed Off Mill having Spiral Teeth.

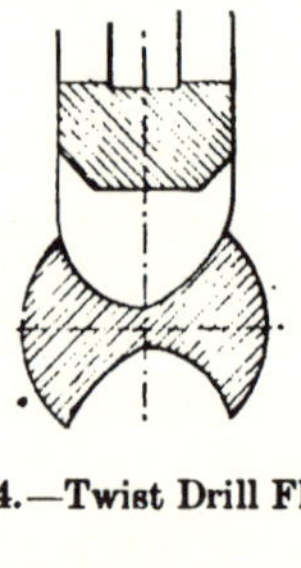

Fig. 164.—Twist Drill Fluting.

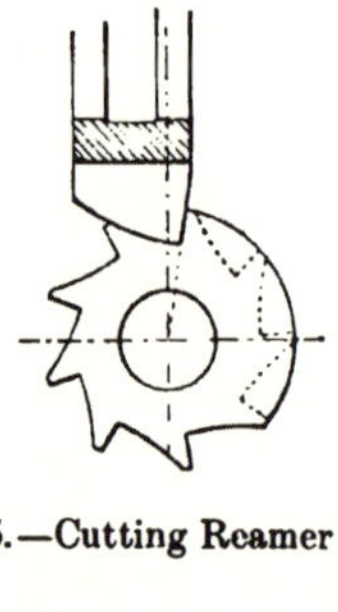

Fig. 165.—Cutting Reamer Flutes.

maintained correctly until the tooth thickness is ground away, by the simple device of striking them from a different centre than the centre of the cutter diameter. This is now adopted, not only in the cutters for wheel teeth, but in those also for milling most profiles of irregular kind. The value of the device in prolonging the life of cutters and ensuring the interchangeability of parts is incalculable.

Figs. 162 and 163 show formed mills being cut, the first with straight, the second with spiral teeth. Afterwards the teeth are backed off in a relieving lathe or in a rig-up on an ordinary lathe. Grinding is always subsequently done on the radial cutting face, never on the backed-off portion, and thus the sectional shape remains unchanged until the teeth are too weak to perform service.

Some special cutters are shown in succeeding figures operating on cutting tools. Fig. 164 shows the milling of a flute of a twist drill, Fig. 165 cutting a concave-toothed reamer, Fig. 166 a convex-toothed reamer or boring tool, and Fig. 167 cutting tap flutes.

Grinding and Sharpening.—To go into the construction of

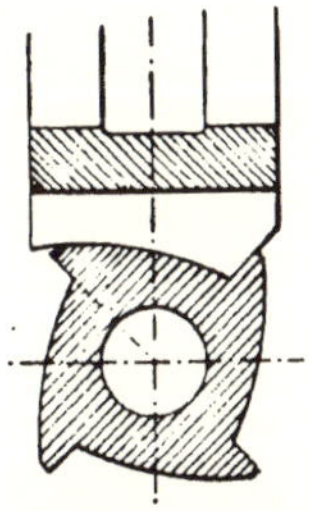

Fig. 166.—Cutting Four-lipped Boring Tool.

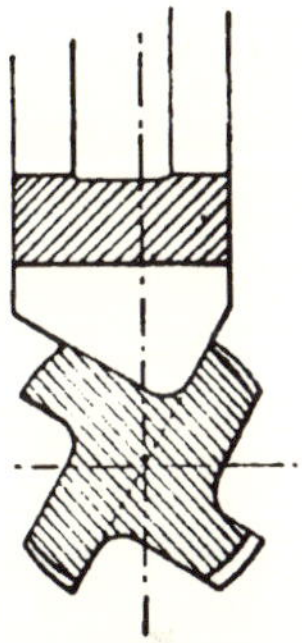

Fig. 167.—Milling Tap Flutes.

the cutter-grinding machines would lead us too far afield. They are numerous, ingenious in design, and the results are arrived at in various ways. We are content to deal with the results achieved. Fig. 168 must suffice for illustration.

This machine provides for universal movements to perform all classes of grinding. The table is carried on a knee, which may be swivelled around the circular column and clamped at any location thereon. The table also swivels, and a head is fitted to the latter having capacity for horizontal and vertical angling, so that the work may be presented to the wheels at any angle. All these angles can be read off on dials, so that their precise amount is known. The table travels to and from the column by means of a

Fig. 168.—Cincinnati Cutter Grinder.

graduated screw, allowing of movements of one-thousandth inch being read off. Adjustable stops limit the travel.

The head carries two wheels—a cup and a disc—which are employed according to their respective suitability for the work in hand. Two speeds are available. The table attachments, besides the angling head mentioned, include a tailstock with a spring plunger allowing for endlong expansion of work, a vice mounted on a swivel face to allow of angular movement being given to the work gripped in the vice jaws, an internal grinding attachment having a small spindle running in long sleeve-bearings and carrying a little wheel at the end. This is driven by a pulley on the spindle actuated from the overhead countershaft. A gear-cutter grinding attachment is fastened to the table, having a vertical stud to carry the cutter by its bore and a pawl to lock each tooth in

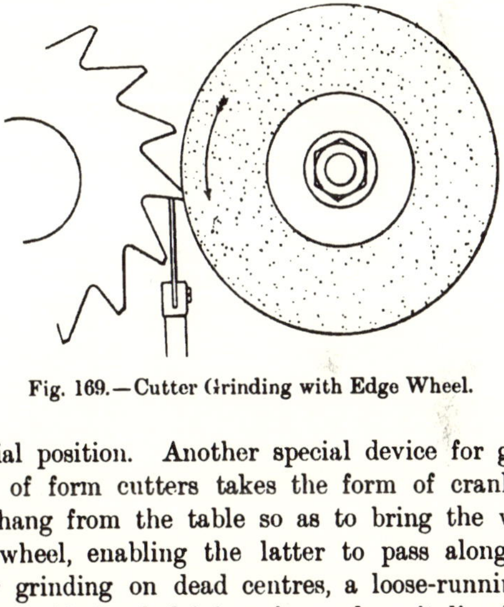

Fig. 169.—Cutter Grinding with Edge Wheel.

correct radial position. Another special device for grinding the tooth faces of form cutters takes the form of cranked centres, which overhang from the table so as to bring the work underneath the wheel, enabling the latter to pass along the radial faces. For grinding on dead centres, a loose-running pulley is fitted to the table-head, driving the work as it lies between the centres. The height of cutters in relation to the wheels is regulated by the table elevating mechanism provided with a micrometer.

The capacity of the machine is 16 inches between centres by 8 inches diameter. Face mills up to 12 inches and saws up to 24 inches can be treated.

Almost invariably cutters are ground and sharpened in one of two ways—either by passing the grinding wheel over the edges of the teeth as in common mills, or along the faces as in formed mills. The practice of sharpening over the entire tooth space (in saw-sharpening fashion) is unusual because wasteful of time and productive of heating.

Taking the edge grinding first, this may be effected by different wheel shapes. The principal results desired after the equal length

Fig. 170.—Grinding with Edge Wheel on a Cincinnati Machine.

of teeth are a correct and uniform clearance angle, and a flat face rather than one concave.

Figs. 169 and 170 illustrate the most common method of grinding axial cutters. The only objection to this is the concavity formed by the wheel. To diminish this, the wheel is selected as large as is convenient, and so long as it does not foul the tooth above, it may be any diameter in reason. In different machines the wheel is presented differently, either in the manner shown, or vertically above. Many prefer to use a cup wheel, Fig. 171, which leaves a flat face to the edge of the tooth.

The tooth rest is an essential fitting, not only for pitching the teeth and supporting the wheel, but also for fixing the angle of clearance. After the rest, in Fig. 169, has been set to suit any one cutter, its position is not changed until the cutter is completed, but each tooth in turn comes round on it. The socket which carries the rest is provided with a wide range of movements to suit plain, cylindrical, spiral, and angular cutters. The blade is often formed of an elastic strip of steel, sometimes solid, sometimes divided as in Fig. 172. The object of dividing it is

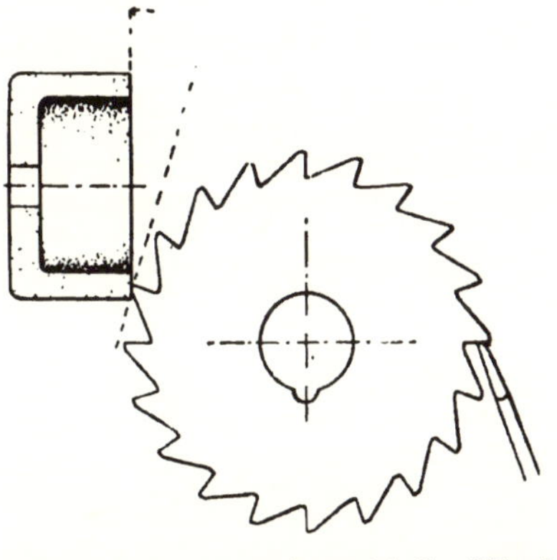

Fig. 171.—Cutter Grinding with Cup Wheel.

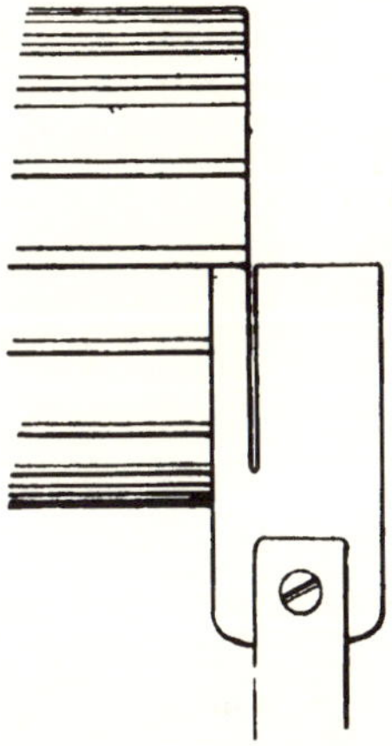

Fig. 172.—Tooth Rest.

that the narrower portion can be sprung out and in from tooth to tooth, without running the rest quite clear, so saving time, and avoiding risk of missing a tooth.

The amounts of clearances on milling cutters are important. They should vary only from about 5° to 7°, the first being suitable for finishing, the second for roughing. The Cincinnati Milling Machine Company gives tables by which these clearances can be obtained by the setting of the tooth rest below the centring gauge. The tables are for cup wheels, and for disc wheels, and are given overleaf.

Cup Wheel Clearance Table.

For setting tooth rest to obtain 5° or 7° clearance when grinding peripheral teeth of milling cutters with cup-shaped wheel.

Diameter of Cutter.	For 5° Clearance.	For 7° Clearance.
Inches.	Inches.	Inches.
$\frac{1}{4}$	·011 = 1-64 −	·015 = 1-64
$\frac{3}{8}$	·015 = 1-64	·022 = 1-64 +
$\frac{1}{2}$	·022 = 1-64 +	·030 = 1-32
$\frac{5}{8}$	·028 = 1-32 −	·037 = 1-32 +
$\frac{3}{4}$	·033 = 1-32	·045 = 3-64
$\frac{7}{8}$	·037 = 1-32 +	·052 = 3-64 +
1	·044 = 3-64	·060 = 1-16
$1\frac{1}{8}$	·050 = 3-64 +	·067 = 1-16 +
$1\frac{1}{4}$	·055 = 1-16 −	·075 = 5-64
$1\frac{1}{2}$	·066 = 1-16	·090 = 3-32
$1\frac{3}{4}$	·077 = 5-64	·105 = 7-64
2	·088 = 3-32 −	·120 = 1-8
$2\frac{1}{4}$	·099 = 3-32 +	·135 = 9-64
$2\frac{1}{2}$	·110 = 7-64	·150 = 5-32
$2\frac{3}{4}$	·121 = 1-8	·165 = 11-64
3	·132 = 1-8 +	·180 = 3-16
$3\frac{1}{4}$	·143 = 9-64	·195 = 13-64
$3\frac{1}{2}$	·154 = 5-32	·210 = 7-32 −
$3\frac{3}{4}$	·165 = 5-32 +	·225 = 7-32 +
4	·176 = 11-64	·240 = 15-64
$4\frac{1}{4}$	·187 = 3-16	·255 = 1-4
$4\frac{1}{2}$	·198 = 13-64 −	·270 = 17-64
$4\frac{3}{4}$	·209 = 13-64 +	·285 = 9-32
5	·220 = 7-32	·300 = 19-64
$5\frac{1}{4}$	·231 = 15-64	·315 = 5-16
$5\frac{1}{2}$	·242 = 1-4 −	·330 = 21-64
$5\frac{3}{4}$	·253 = 1-4	·345 = 11-32
6	·264 = 17-64	·360 = 23-64
$6\frac{1}{2}$	·286 = 9-32	·390 = 25-64
7	·308 = 5-16	·420 = 27-64
$7\frac{1}{2}$	·330 = 21-64	·450 = 29-64
8	·352 = 23-64	·480 = 31-64

Disc Wheel Clearance Table.

Giving distance for setting centres below centre of spindle to obtain 5° or 7° clearance with wheels of different diameters when grinding with periphery of disc wheel.

Diameter of Emery Wheel.	For 5° Clearance.	For 7° Clearance.	Diameter of Emer Wheel.y	For 5° Clearance.	For 7° Clearance.
Inches.	Inches.	Inches.	Inches.	Inches.	Inches.
2	3-32	1-8	4¼	3-16	17-64
2¼	3-32+	9-64	4½	13-64	9-32
2½	7-64	5-32	4¾	13-64+	19-64
2¾	1-8	11-64	5	7-32	5-16
3	1-8+	3-16	5¼	15-64	21-64
3¼	9-64	13-64	5½	15-64+	11-32
3½	5-32	7-32	5¾	1-4	23-64
3¾	5-32+	15-64	6	17-64	3-8
4	11-64	1-4			

Note.—If emery wheel selected is too large, that is, if it scores the next tooth, a smaller wheel should be chosen and the centres readjusted so as to be right for this wheel.

It makes no difference if the mill is spiral. In the latter during its traverse, the teeth bed upon the rest, Fig. 173, and the

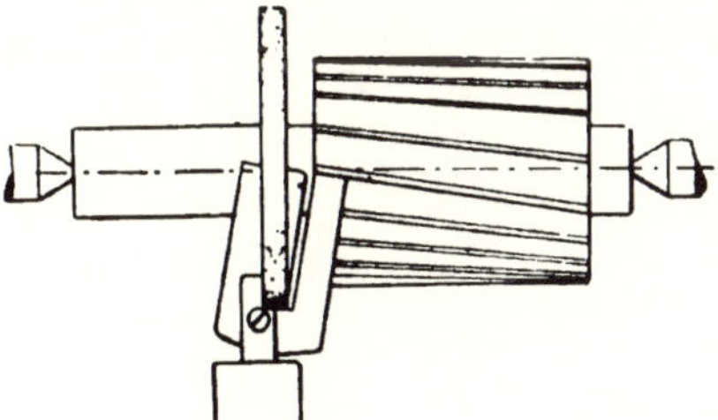

Fig. 173.—Tooth Rest set for Spiral Cutter.

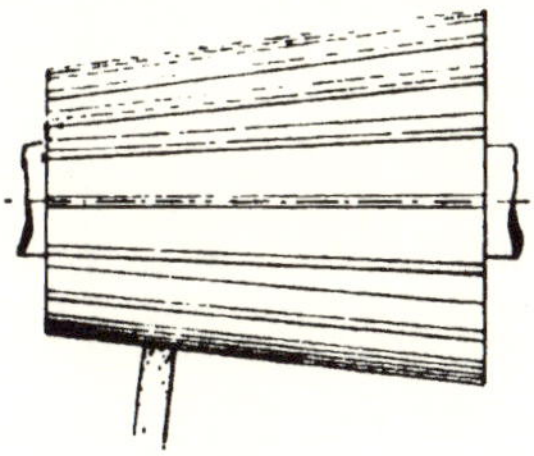

Fig. 174.
Grinding Tapered Cutter.

portion of the length of the spiral which is being ground at any instant is passing the cutter at a constant height, which is fixed by the setting of the rest. The rest is also used for staggered mills and hobs. The teeth of angular cutters also receive support from tooth rests. Taper mills are treated similarly, the axis of the cutter is set to half the taper; and edge or cup wheels are used, Fig. 174.

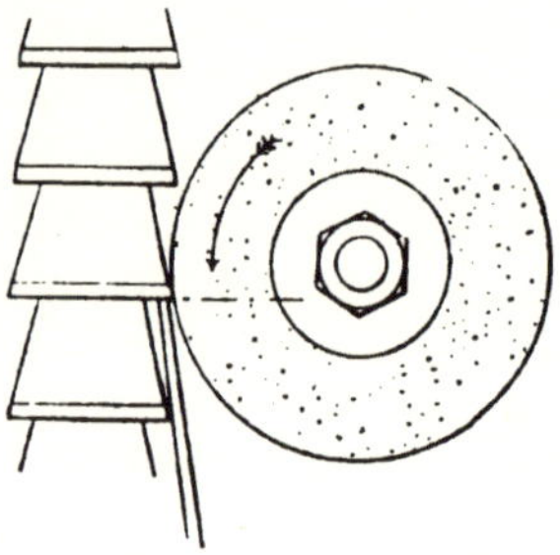

Fig. 175. Grinding Side Teeth.

The side teeth of cutters are ground with disc cutters, Fig. 175, with a wheel as large as possible without fouling the tooth above. Or frequently the wheel is set vertically over the cutter, with the axis of the latter vertical instead of in a horizontal plane. Or cup wheels are used alternatively, as shown in Figs. 176 and 177, the cutters being set for angle

Fig. 176.—Grinding Side Teeth with Cup Wheel.

of clearance in each case, but the wheels being differently presented.

Fig. 177.—Grinding Side Teeth with Cup Wheel.

The teeth of end mills are ground with cup wheels, Fig. 178, the wheel axis, or that of the cutter, is set out of parallel by the amount required to give clearance to the teeth. Fig. 179 shows the hole of a cutter being ground with the internal attachment.

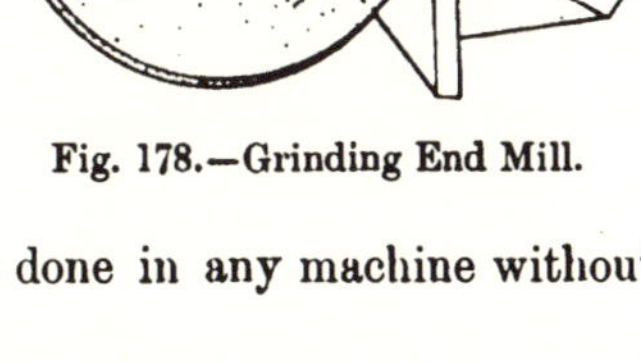

Fig. 178.—Grinding End Mill.

Profile mills if ground on their edges are done by the aid of a former. An example is given in Fig. 180. In this class of work the relieved cutters are generally to be preferred, because being ground on their faces, they can be done in any machine without rigging-up a profile arrangement.

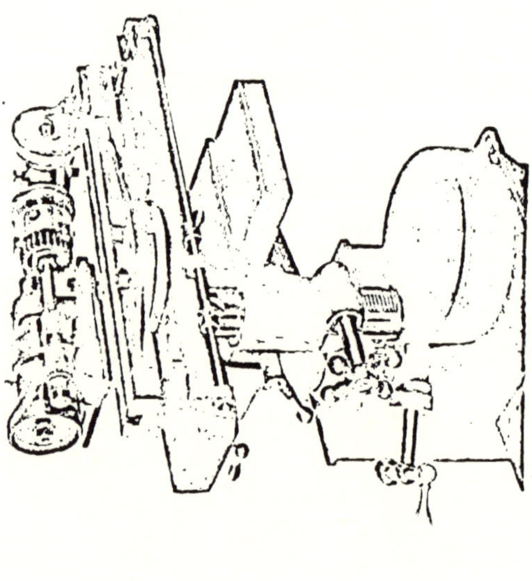

Fig. 179.—Grinding the Bore of a Cutter.

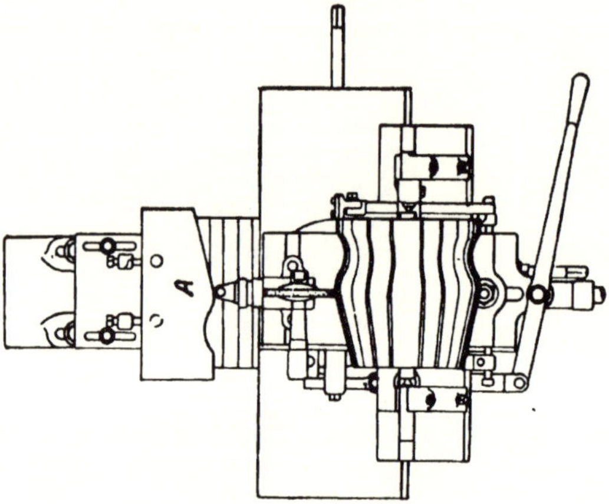

Fig. 180.—Grinding Profiled Cutter.

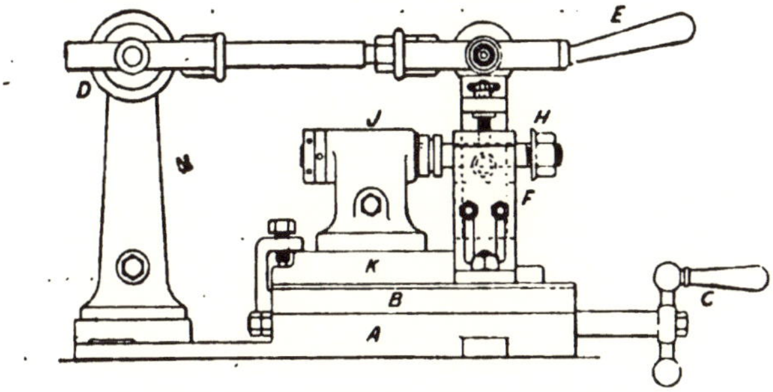

Fig. 181.—Elevation of Form Cutter Grinder.

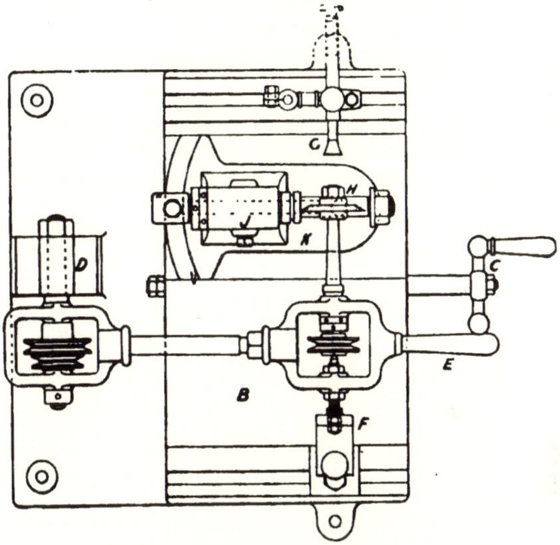

Fig. 182.—Plan of Form Cutter Grinder.

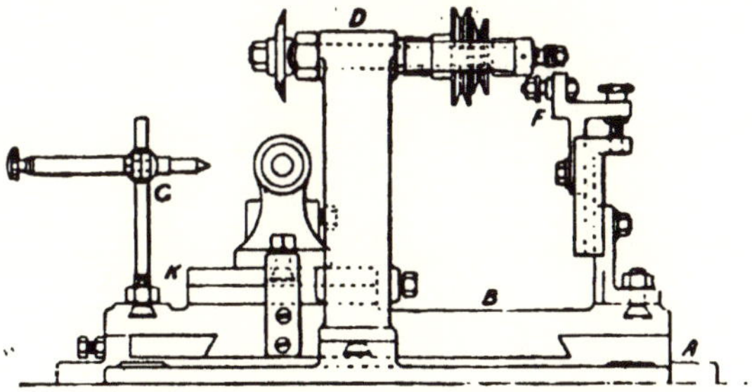

Fig. 183.—End View of Form Cutter Grinder.

Figs. 181-183 illustrate a form cutter grinder by the Actien Gesellschaft für Schmirgel ünd Machinen Fabrikation, of Bockenheim. A is the base, bolted to a bench or a pillar. On it the sliding table B is capable of adjustment by the handle C, operating a screw. The emery wheel carriage is pivoted at D, and is lifted, lowered, and controlled by the handle E. The form or profile piece is bolted to the bracket F, made in three pieces, giving adjustments in the vertical direction,

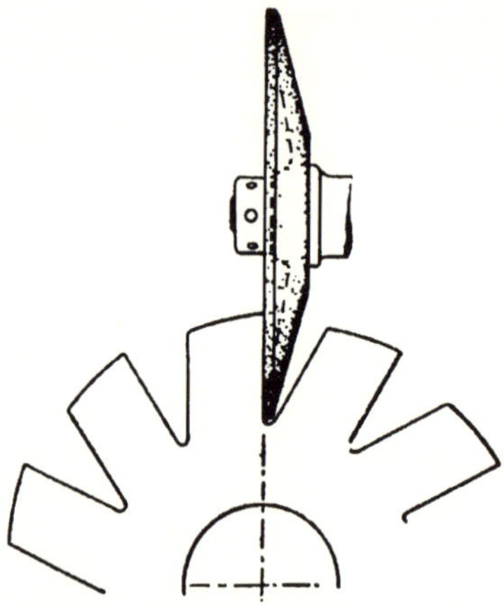

Fig. 184.—Grinding Formed Cutter.

and horizontally, on a sliding face on the table B. The tooth rest G, on the opposite side, has vertical adjustments, and two horizontal ones at right angles with each other. The work spindle H is held in a head J, having endlong adjustments on the base K, which swivels on the table B, upon which it can be clamped. The wheel has two speeds by coned pulleys, the tension being adjusted by a screw and nut in the wheel arm.

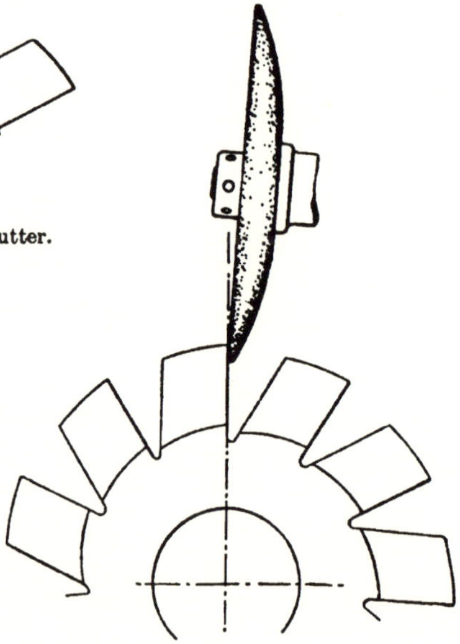

Fig. 185.—Grinding Formed Cutter.

The second method of grinding adopted, that of faces for form mills, is simple. The shapes of the wheels used and their presentations are illustrated in Figs. 184 and 185. The resemblance to that of grinding a tap is shown in Fig. 186.

The economy of regrinding is too valuable to be neglected. It is not only that better work is produced by keeping cutters sharp, but that the operation of grinding is also more readily done. Heavy grinding produces heating, and distortion, and draws the temper, as indicated by blueing. If a cutter has been treated badly by being allowed to become very dull, it is better to regrind it while revolving on a mandrel, before taking it tooth by tooth. Dry grinding is usually adopted. In some firms the cutters are sharpened more finely by rubbing each tooth with an oil stone, or with an emery rubbing stone, or a carborundum stick. Firms who adopt the practice speak highly of the results.

Fig. 186.
Sharpening Tap.

CHAPTER VIII.

MILLING OPERATIONS.

The Operations of Roughing, Finishing, and Profiling—Milling Compared with Planing, Shaping, and Slotting—Examples—Holding Work in Jigs—General Considerations—Examples.

The Operations of Roughing, Finishing, and Profiling.—The time has now come when the true position of the milling machine in the engineers' shop may be considered as practically determined, at least for a long term of years. We have seen that its utilisation in engineers' work did not follow until half a century subsequent to its invention, its applications during that period lying in the small-arms and sewing-machine industries. Then, for many years, engineers used it for light operations only, and chiefly in those which had been hitherto performed at the vice. Milling did not come into much rivalry with the general operations of the machine shop until within about the last ten years. An exception must be made in the case of slabbing face-milling cutters, with inserted teeth, which have been used for twenty years or more, in some shops for rough facing.

About ten years ago the milling machine began to boom in general engineers' work, and it made its appearance in many shops where it had hitherto not been seen at all. Then something like a reaction occurred, and many thought that the multiple cutter was going to displace the single cutter on planer, shaper, and slotter. Practical mechanics did not, as a rule, share that delusion. What has happened is, that though some displacement has occurred, the milling machine has taken charge of a comparatively new field of work, the nature of which we propose now to consider briefly.

From this point of view we may regard milling in three broad aspects—that of roughing, finishing, and profiling.

The first two divisions correspond in the main with two broad types of cutters—the inserted tooth type, and the ordinary continuous edged cutters. Under the first are included the staggered mills. The difference in working between the two broad types is, that the first are capable of slogging, the latter are not nearly so well adapted to that operation. The first comprise a series of single-edged tools, the second resemble the float-cut file in their action. The first are frequently true cutting tools, with front rake, the second are rarely so, since their cutting faces are radial, or normal to the surface being cut. These broad differences between separate and continuous edges, and between cutting and scraping tools lies at the basis of the distinction between roughing and finishing operations.

But roughing is a term of relative meaning, denoting generally the first, or often the second cuts on a job, as preparatory to the final or finishing cut. A cut may be a roughing one without being deep, but a finishing one is always fine. Both are generally done with the same mill, fed deeper, or shallower. But no cut taken with continuous edged mills can ever be as deep as that which is taken with a single-edged tool, or with staggered mills. The difference is that between scraping and cutting, between broken shavings and coarse chips. It is because this distinction has been often neglected that too much has been expected from milling cutters.

Limits are set to the action of single-edged tools in lathe, planer, shaper, and slotter by the spring of rests, of cross rails, and rams. The stresses on a broad-cutting finishing tool prevent any but very shallow cuts being taken. Yet milling cutters exceed in width many times those used for broad finishing, and therefore so much less depth of cut can be expected of them. This increased breadth of action, the breaking up of the chips by the scraping process, and another fact, the small spacing of the teeth, tending to cause choking and dragging, are the disadvantages under which the milling cutter does its work.

What has partly disguised these facts is the abundant lubrication imparted to these cutters in the best practice, which contrasts strongly with that of the single-edged tools; the wider spacing of the teeth given for some kinds of work, the more efficient support afforded to cutters and work than of old, and the spiral arrangement of teeth, which effects a shearing cut. Features which may conduce to a better solution of the difficulty are the adoption of a more rational method of feeding, and a further stiffening up of the machine elements.

The roughing operations of the milling cutter may simply mean the first, or the second cuts of the common mill, preparatory to the finishing cut of the same tool. Or it may mean roughing performed preparatory to planing, either by a common mill, or gang of cutters, or by means of an inserted toothed mill, or of a staggered toothed solid mill. Or it may mean rough tooling only, as in that large class of face cutters which are used for ending rolled beams, or facing the feet of standards or brackets, or other castings in which the object is to remove the rough exterior, and leave a roughly faced surface, suitable for attachment to other parts. Each of these is largely represented in the practice of the shops. The finishing operations of the miller include the greater portion of the work done on these machines, and these can never be any but very shallow cuts, consistently with accurate results, and the wider the surface, the less must be the depth tooled. So that practically we may say that broad and accurate finishing is not where the miller scores best, though it is possible under the conditions just named. Its best work is done on narrow pieces, or on those of medium width only.

Milling Compared with Planing, Shaping, and Slotting.—Milling over a broad surface or on a number of narrow pieces arranged side by side having their aggregate area covered with a single broad cutter is a rival operation to the single tool, or couple of tools held in the tool box of a planing machine. When the question is one of roughing down, the milling machine scores; when it is that of very accurate finishing, the planing machine is found superior in accuracy, though not perhaps in smoothness of surface. The question, therefore, is not an absolute one, but relative. And the reason is fairly clear to a practical hand, being due to the difference between the spring of a broad cutter operating over a width, say, of 12 to 20 inches, and the rigidity of a shanked tool taking a narrow cut. It is practically impossible to avoid spring on any cutting that is both heavy and broad in character. Spindles are being constantly stiffened up, more attention is given to their bearings; cutters are serrated and staggered for breaking up the chips, and provided with teeth having a high angle of spiral for finishing, yet they spring and chatter. And even though the amount is not much, it is sufficient to destroy the very fine accuracy that is required in machine tool slides, and in other intimately fitting parts. The conditions are not comparable with those of a narrow milling cutter on a stiff arbor. Lathe men will understand this when they remember that heavy cutting and very accurate results are incompatible; that fine precision of results requires light cuts, and that it is better to rough coarsely and finish finely than to reverse the process.

The mills with spiral teeth are better for finishing, since they take extremely shallow cuts. But what they lose in depth they may gain in area, and thus show considerable economies over planing: hence their special value for preliminary operations.

In fine finishing, no less than in roughing down, a stiffly built machine, whether planer, miller, grinder, or lathe, is essential. It may sometimes seem incongruous to see a massive machine weighing several tons revolving a tiny milling cutter or emery wheel, nibbling at a bit of work, removing perhaps a hundredth or a thousandth of an inch of metal. Apart from experience, this disproportion would appear absurd. This has not been arrived at by calculation, but it is a case of evolution in the workshop. When the effects of vibration, or slackness of spindle and slides, or even of the spring of parts that are too slender to hold the work or the cutter stiffly, are found to occur in the smallest work, it is not difficult to see what may happen in the case of a very large cutter. The faintest trace of vibration, of yielding of frames and spindles, produces unevenness in the surface of the work, which is fatal to the finer degrees of accuracy required in the production of high-class machinery: hence machines are constantly being built of more massive design than their predecessors.

Plain surface tooling—using mills which cut by their edges—constitutes the largest class of work done. For this the mills should be longer than the width of the surface which has to be operated on. They are either vertical or horizontal. The work may be a single piece, or a number of pieces arranged in series. Mills of this general type are used on every class of machine, and on an infinity of duties. Mills about an inch in width have the teeth arranged spirally to give a shearing cut, which separates the metal in detail, and puts less strain on the arbor. The amount given varies widely by different firms, but in recent years the angle of spiral is greater than was formerly adopted.

When employing broad cutters, a great deal of work is either ground or planed after being milled, being so treated in order to impart a higher degree of accuracy than is found practicable under ordinary conditions when using broad cutters in the milling machines. When work leaves a planing or shaping machine it is finished, except when scraping for high-class work only is required. The practice just instanced seems a proof that broad, heavy milling, with axial mills at least, is not yet so accurate as planing. There are exceptions, but they serve to accentuate the rule that the milling of heavy work with deep cuts and rapid feeds is chiefly valuable in removing the bulk of the material and producing a fairly accurate surface, which then has to be finished by light grinding or planing. Much of this is probably due to the employment of light machines on work for which they are not adapted, and to the practice of performing too heavy cutting on large areas, with arbors also too light for the work.

Wide faces and inner edges can be tooled with broad face and side mills, many examples of which occur in the slides of machine tools. Smaller flanking mills may cut edges simultaneously. The angular mill affords good facilities for tooling the angles of slides accurately.

Face-milling cutters are preferable for some purposes to cylindrical cutters, since they are useful for depthing, as well as for surface work.

Another advantage which the large face mills with inserted teeth possess over single-edged planer tools is, that these will reach down or across to surfaces that cannot be reached with a planer tool, properly supported. There is a great difference between the overhang of the shank of such a tool from the tool box, and the mass of a cutter head.

A limitation of the milling machine is that a mill cutting axially must always leave a radius in corners of recessed portions of work. If this is objectionable, it will often happen that a face mill can be substituted. But in many cases a radius is to be preferred. The following are some of the principal jobs done by milling, several examples of which are shown on subsequent pages:—

The flanges of steam chests can be tooled by a face mill on a horizontal spindle, and the truth of the work will be ensured by bolting one end flange, already faced when boring, on the machine table. At the same setting the ports can be milled.

Three sides of parallel bars can be tooled simultaneously with rotary mills on a machine of the planer type, fitted with one vertical and two horizontal heads. Or two edges can be milled simultaneously with a machine having two vertical spindles only.

The shoulders between brasses are shaped parallel, with mills having teeth on ends and edges. All kinds of brass fittings have their angular faces tooled in special or plain machines. Bearing caps have their edges milled with edge mills on vertical spindles. The faces of eccentric straps are tooled with edge and with end mills. So are keys and cottars, slide valves, and eccentric rod links. The edges of cycle cranks are tooled with profile cutters. Taps, reamers, and twist drills are grooved, as already shown, between centres with cutters having the profiles of the grooves. Tee grooves and vee grooves are milled with cutters of corresponding sections carried in vertical or horizontal spindles. Vee ways for slides of moderate width can be tooled at once with a one solid mill. The teeth of racks are milled, several at a time, by a gang of cutters. The rims of gear wheel blanks are milled to section on periphery and edges, the blank being fed circularly on a horizontal spindle. The square ends of spindles can be cut with end mills. Vice jaws can be milled, and oil channels in small tables. Slits for tightening purposes are cut with slitting saws on a milling machine instead of casting them. Then there is besides the work of cutting all kinds of gear wheels, which subject will be treated in another chapter.

Key ways and that class of grooves can be tooled with mills cutting at bottom only, or at bottom and edges at once, the shaft or other piece of work being clamped to the table or in the vice. The tool is a disc with teeth on the periphery only in the narrower sizes, but with teeth also on the sides in the wider ones. In the narrow ones the teeth are square across, in the wider a spiral twist is given to them.

Faces and outer edges can be tooled simultaneously by what are termed straddle mills, and a constant width maintained easily in a large number of pieces. By turning the side mills round, the outer edges can be used after the inner ones have become dulled. Also, edges can be tooled thus without touching the face by inserting a plain distance piece between side mills. Or a single edge can be tooled with a side mill of sufficient size to cover the area required.

A convex edge is a very simple detail. These edges occur with great frequency in engineers' work. The old method was to plane or shape them, adjusting the tool slides by hand, or to file them. The modern method is to mill them with concave cutters if half-circles, or with cornering or radius cutters if quarter-circles. These can be used equally well on vertical or horizontal arbors. Concave forms can be milled.

Half-bearings or keeps can be milled singly or in series, instead of boring.

More elaborate sectional forms can be cut by mills operating by edges and ends: and these may run in one plane or irregularly, as in profile work. They may be formed in one piece or built up in gangs. There is no limit to such combinations, save the strength of arbors and the capacity of machines.

Large profile cutters are multiplied in gangs on an arbor to tool several sections at once. The ways of small lathe and machine beds are thus shaped. These gangs are made up more cheaply than by forming solid mills for intricate work, besides which the separate mills can be used for plain work. They may lie closely together, with or without overlapping of teeth, or plain distance collars may be used to separate them, if the cutting is not continuous. Edge mills may also fulfil the function of distance pieces. Mills are locked together in gangs by projections on one entering recesses in the one adjacent. Several similar mills may be arranged side by side for duplicating work. In cases where an exact dimension across faces has to be worked by, two mills may cut on each face at once, whether the faces are internal or external. Where the alteration in size of a single mill is objectionable because the later pieces tooled would not be of the same dimensions as those first done, then, for inner faces, interlocking mills are used, and for outer ones mills separated by a collar, either device of which permits of micrometric readjustment.

The utilities of the miller are very great on form, and on profiled work. The distinction made is that between the work of the gang mill operating on a plain surface longitudinally, and that of any edge mill, gang or otherwise, operating on a surface which is irregular in the plane of the axis of the mill. There is no other tool in the machine shop which will produce such surfaces. Before the mill came they were done by hand at the vice, or in a rough fashion by a succession of settings of tools and cuts on the planer, shaper, and slotter. On the milling machine they can not only be done cheaply, but with nearly absolute uniformity, which was not possible under the old system. This, therefore, is a new sphere which the miller has almost wholly created for itself.

There is another aspect of the methods which either came in with the milling machine, or the development of which it has greatly helped. Since cutting, to be accurate, must be shallow, and since slight allowances are only consistent with very close approximation to dimensions, the practice of milling has reacted on the work of the die forger, and of machine moulding. As these methods are only economically available when considerable numbers of similar pieces are required, they have favoured the development of interchangeable methods, and the practice of jig making. The latter has reacted on the work of the smith and moulder, in demanding the closest adherence to dimensions, and the smoothest surfaces possible, because it is difficult to design jigs for pieces of work that vary in size and form, and it is wasteful of time to have to set such pieces in jigs. And as allowances diminish, the presence of scale becomes more objectionable than when a good roughing cut can be taken with a single-edged tool, penetrating below it at once. Pickling therefore is resorted to, both for castings and forgings, in shops where consideration is given to the permanence of the edges of cutters.

The relative degree of accuracy of surfaces which is desirable or essential must frequently control the decision of the alternative methods of planing or milling. There is, for example, no comparison practicable between milling a number of girder ends, or stretcher ends, or pipe flanges, and milling a long lathe bed, or planing machine bed, or slides. The two classes of work, and the large group of which each is broadly representative, stand in an entirely different category. Lathe and machine beds are not considered true enough unless they are within about $\frac{1}{1000}$ of an inch in a length of 6 feet. Now an extremely slight amount of deflection under cutting will produce this minute degree of inaccuracy. But for a large proportion of faced work, $\frac{1}{100}$ of an inch would be sufficiently accurate.

Then further, the question of whether roughing or finishing is being done is on a par with that of relative degree of accuracy. If a piece of work has to be finished by grinding, then fine accuracy in the roughing is not important, but the roughing may correspond with the degree of accuracy, say $\frac{1}{100}$ of an inch, which might be the finish limit in a different grade of work.

The springing of work under cutting is a matter of which account must be taken when comparing alternative methods of tooling, and deep or shallow cutting, coarse or fine feeds. It may be accepted as beyond doubt that all planing or milling causes some deflection and springing, though the amount would be very variable under different conditions. But its presence and amount would in numerous instances be the principal factor in deciding the choice between planing and milling, or between roughing by one method and finishing by the other, or between taking heavy and light cuts, or between using inserted tooth rotary face mills or solid edge mills.

The question of milling *versus* planing lathe beds has been and is to an extent an open one. In some cases it is solved by employing milling as a roughing operation, and planing as a finishing one. But planing involves the employment of a gauge or gauges as the work proceeds, and several settings of the tools. Hence the idea of a single gang of mills tooling the shears at one operation has its fascinations. But such work is only practicable under certain conditions.

The great difficulty encountered is that of spring, both that of the gang of cutters, and that of the beds, and these forbid the employment of any but very shallow cuts for broad finishing. A single-edged tool, though heavily fed, can be trusted not to cause spring in a bed, but the case is on an entirely different footing in a gang of cutters operating over a width of several inches.

This elasticity may be minimised, but not eliminated entirely. Stiff arbors are essential, and careful levelling and packing up of the beds with wedges, or with special appliances, as stools and clamps in a fixture attached to the machine bed. Clamps must also always be pinched right over the points of support that are employed. After a rough cut has been taken, the clamps should be slackened to note if the bed has sprung in consequence of the removal of the skin, which is liable to happen in the lighter class of castings. Also light cuts taken over the rough skin of a casting ruin milling cutters, so that between these and the heavy ones, which are impracticable, there is little if anything to be gained by milling, especially when extra expense may be entailed for final scraping.

The work however may be divided with advantage, either by roughing down with single-edged planing tools on the planer, and finishing with very light cuts by a gang of cutters operating over the whole width. This involves setting at two machines, unless a combined planer and plano-miller is available. Or the roughing down may be done by mills, roughing sections of the work only, with coarse feeds, to be finished as before with a broad gang mill. Given machines specially rigid in beds, tables, arbors, and cross slides, there is no reason why milling should not be both broad and accurate in character. Thus, at Messrs L. Loewe & Co.'s, lathe beds are roughed out on the milling machine, and finished on the planer, the latter being employed to ensure greater accuracy. The precaution is taken of allowing the work to stand for several weeks, after rough milling, before it is finished by planing, to give time for the slight change of shape that occurs to take place.

Screw machine beds are regularly milled in the Warner & Swasey shops with gang mills on a planer type of machine. The gang mill goes over the surfaces of the vees, the bed edges, and the rack bed. Supplementary spindles in special heads, with a slight vertical range of adjustment, mill the bottom faces for the gibs, using shank mills. Roughing and finishing cuts are taken thus. The beds are finally corrected by scraping.

The problem of milling *versus* planing is not only governed by the spring of the work, and the relative quantities of metal removed, but by the relative facilities existing for tooling faces at different angles and in different planes. The planer is the more accommodating of the two in this respect, because tools can be cranked and set to operate on undercut faces, often in situations that neither vertical nor horizontal cutter spindles are able to reach. In consequence of this, numerous jobs can be tooled at a single setting on the planer that would require two on the milling machine. And some could not be finished at all on the latter, or if so, an expensive special tool would have to be made, and special tools are not to be considered in the work of the general shops.

Face, and edge milling afford two alternatives which often have to be discussed and considered in relation to special jobs. The question of rotary planing is also often included when face milling is concerned. In a good many cases there is no alternative, as when faces on the same job lie, some horizontally, some vertically; or as when a job can only be fixed one way on the machine. But when choice can be made, the face mill has the advantage that the chips fall away from the work at once. In edge milling the action of the cutter is impeded by the presence of the chips, which do not get away freely from between the teeth, and this acts as a limitation on the depth of cut and feed. And when castings are not pickled, the sand and scab lying in the path of the teeth aggravate results, so that the horizontal face mill scores more heavily still than when working on clean surfaces. For the roughest class of milling, therefore, the face mill on a horizontal spindle is the best rival to the planing machine.

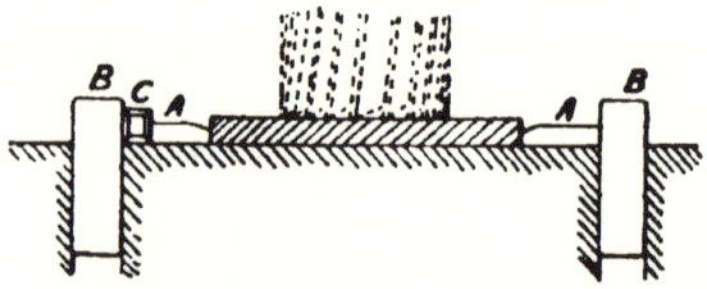

Fig. 187.—End Mill working on Thin Piece held with Side Strips wedged against Round Pins put in Holes in Table.

Some examples of milling work which would alternatively go on the planer are shown in subsequent figures. Taking thin pieces, there are two or three ways of holding, as follows:—A few parallel pieces, Fig. 187, A, of various widths and lengths, chamfered off to a narrow edge down one side and planed parallel, are kept handy to suit different jobs. The thin edges go against the sides of the work, and, being shallow, do not interfere with the milling of the top face. The shorter pieces are used for short bars, the longer ones for long bars. They may be held in place by means of wedges simply, and used in pairs, as in Fig. 187. The holes in the Fig. or the tee slots in a machine table run pretty close together, and, by selecting parallel pieces of different widths to suit varying sizes of bar being tooled, any width can be held on the table. In Fig. 187 plain pins B, inserted in one set of holes, take the thrust of one parallel piece, while the other piece is forced against the bar by means of wedges C, C driven against pins B in other holes.

Another way is to hold one of the parallel pieces A down with a bolt, which may be long or short, Fig. 188, and the other B

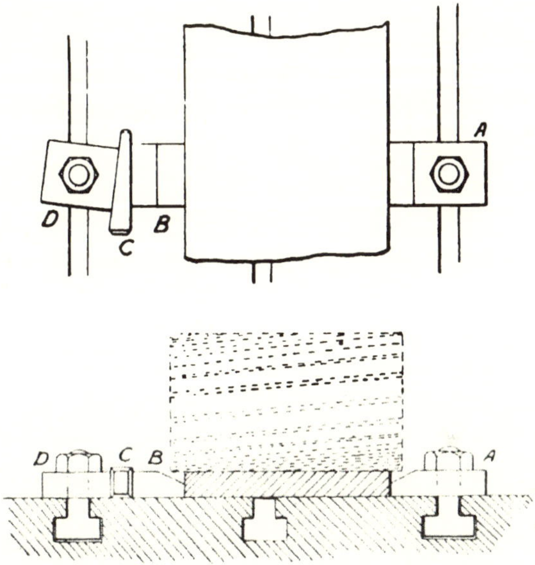

Fig. 188.—Edge Mill working on Thin Piece held with Side Strips wedged against Blocks Bolted in the Slots.

wedged C against a pivoted block D. Or pinching is done by means of adjustable points or fingers and screws. The fingers are bits of round rod, flattened at the ends which bear against the work, and they are hollowed at the other end to receive the push of the tightening screws. The screws are tapped into stops, which, fitting into the table groove, take the reaction due to the tightening up of the screws. Or a bracket of larger size, with feet, can be used, and tightened down with bolts

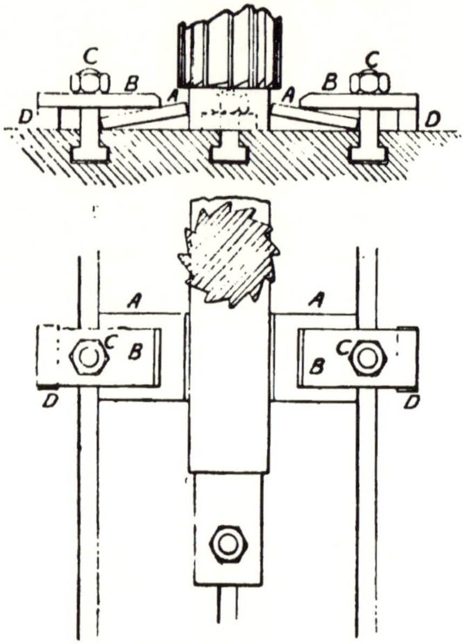

Fig. 189.—End Mill working on Piece held with Clamping Plates and End Stop.

fitting in the table slots. Long parallel guides are further useful in imparting steadiness to narrow, weak bars by affording them some measure of support. Being also parallel, they serve to set the work true with the tee slots or holes.

Bars of rectangular section often have to be milled on all four faces, as, for example, the wrought-iron guide bars used for hydraulic rams, and also those for some engines. In such cases no clips can be used for holding down the bars on the table. There are no side flanges either to afford a bearing for plates. Such work is held by end or by side pressure alone. Then one of the methods shown in Figs. 189 and 190 is adopted.

Fig. 189 shows a long guide bar for a hydraulic ram. The actual clips by which the bar is held are shown at A. These, it will be observed, are pressed against the sides by leverage. They stand at an angle, and are jammed between the sides of the bar and the tee-

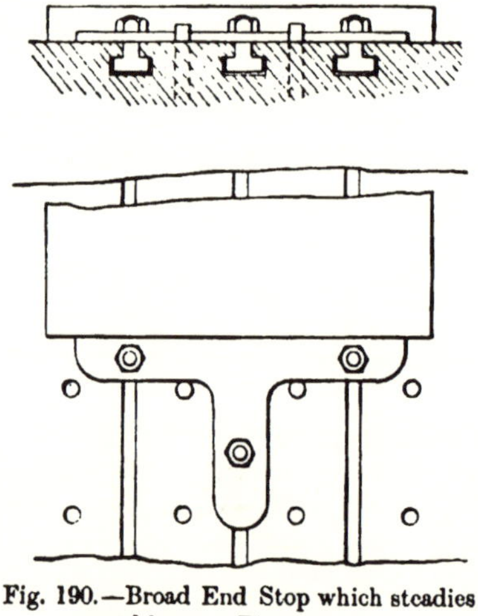

Fig. 190.—Broad End Stop which steadies and keeps a Piece square.

head bolts. The tightening of the bolts C presses down the plates B upon A, the plates B being supported at the outer face upon the packing pieces D. As the plates A are longer than the parallel distance between the sides of the bolts C and of the bar, it is obvious that the tightening of the bolts C must wedge them against the bar, and the more tightly C C are turned the more securely will A A be pinched against the sides of the bar, so that it is held perfectly rigid, and the top face is left clear for milling.

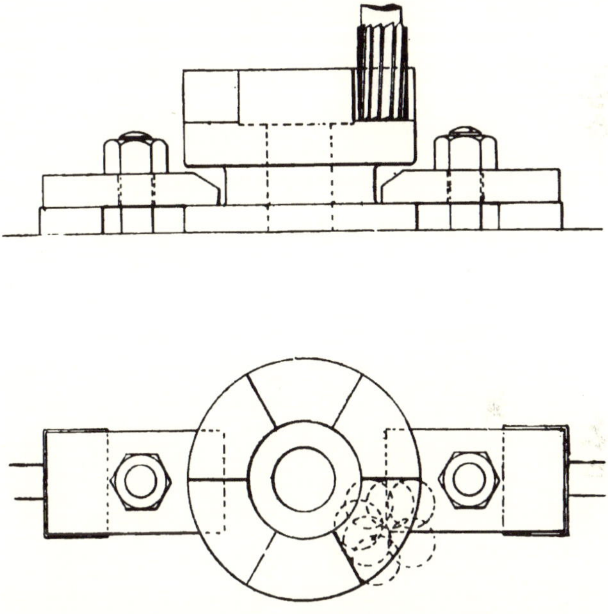

Fig. 191.—End Mill operating on Claw Clutch clamped by its Collar.

The number of bolts and plates will depend on the length of the bar being tooled. There may be four, six, or even eight sets, their distances apart ranging from about 2 feet to 2 feet 6 inches. A stop block is often placed at the end, as seen.

In some light work even, in which it would be practicable to effect a hold with clips upon the top face, it is better to adopt the method illustrated in Fig. 189, because the flank pressure obtained thus is not nearly so liable to pull the work out of truth as the pressure exercised by the direct pinching down of a clip on a light bar, or forging, or casting.

Another way of arranging clips of this kind is to place them at the ends, or a broad steady piece like Fig. 190 may be used. It is suitable for work of no very great length, say, not exceeding

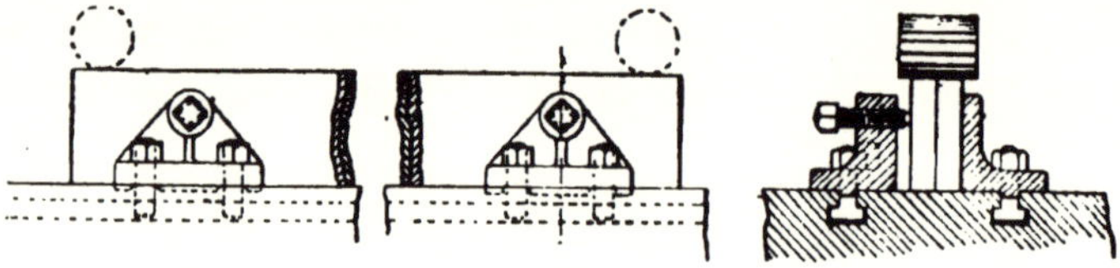

Fig. 192.—Edge Mill operating on Two Thin Strips clamped between Angle Brackets.

3 feet. Fig. 191 illustrates another way of gripping with clips, taking a bearing on a recessed portion—a claw clutch, in this case, but equally adaptable to straight pieces having recessed portions.

Bars may be held on edge by the rig-up shown in Fig. 192, by means of brackets held with tee-head bolts, one set of brackets carrying pinching screws.

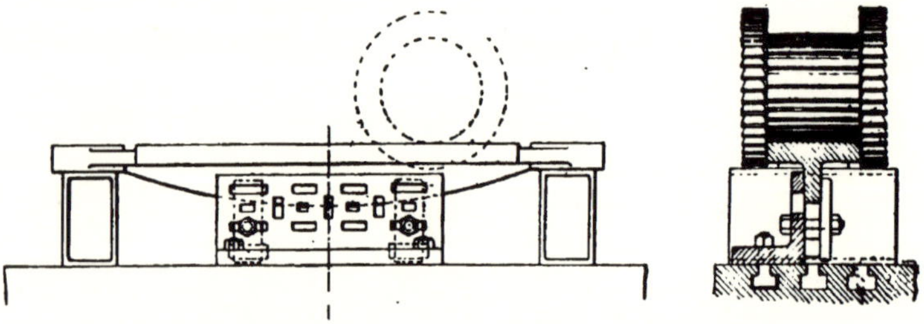

Fig. 193.—Motion Bar, tooled on Face and Edges with Straddle Mill, while bolted against Angle Plate, and supported on Packing Blocks.

Fig. 193 shows how cast-iron guide bars may be held for milling. No clips can be pinched on the top faces of the guide bars because they have to be machined all over, and neither are there holes of any kind which can be utilised as means of holding down. The rib affords the only convenient means of attachment.

Fig. 194 shows the fixing of the curved side plates for tanks, going at the bottom and side of the tank, illustrating the setting of the plate A for milling the longitudinal edges. It is laid

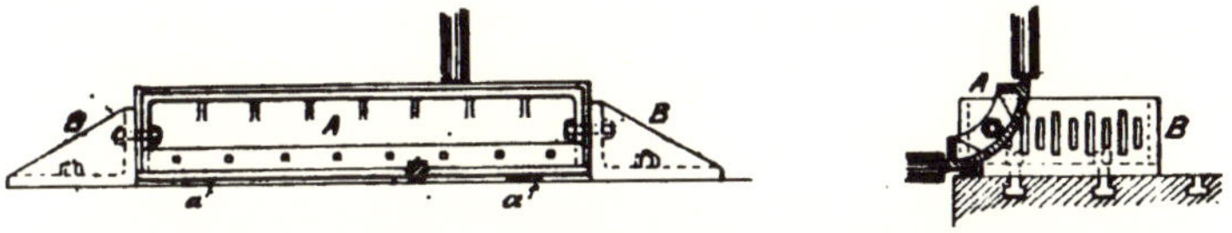

Fig. 194.—Corner Tank Plate, having Joint Faces tooled with End Mills while Clamped, and adjusted between Angle Plates.

directly upon the table of the machine, or on shallow packings *a, a,* and set and bolted to angle plates B, B, which are bolted to the table. These plates always form quadrants of circles. The caulking strips down the longitudinal edges are therefore at right

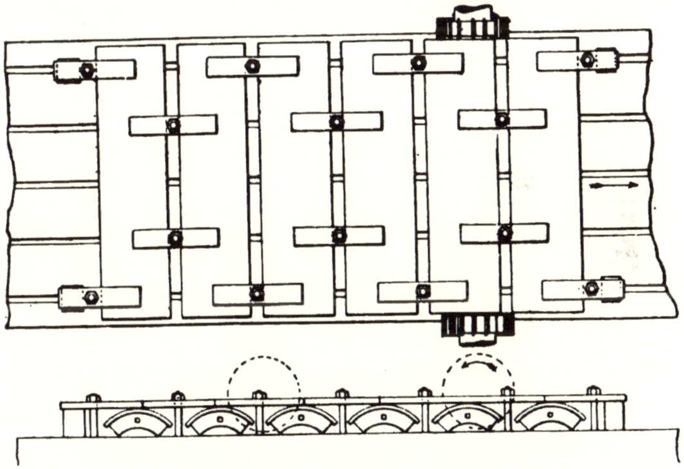

Fig. 195.—End Joints of Corner Tank Plates being tooled with End Mills while held down with Clamping Bolts and Plates.

angles with each other, and the lines of the right angle, when prolonged, meet at the centre from which the radius of the curve of the plate is struck. To set the plates, therefore, they are first lightly bolted to the angle plates when brought nearly into position, and then adjusted exactly with the square and rule. The square is held against one strip or "fillet," and measurement is taken with the rule to the outside of the curve. When this distance—usually either 6 inches or 9 inches—equals the radius of the curve, plus the allowance for tooling on the lower strip, then the plate is square, and the bolts can be tightened ready for milling the strips. Both top and bottom strips are tooled at one setting.

To mill the ends, several plates are laid in line across the table

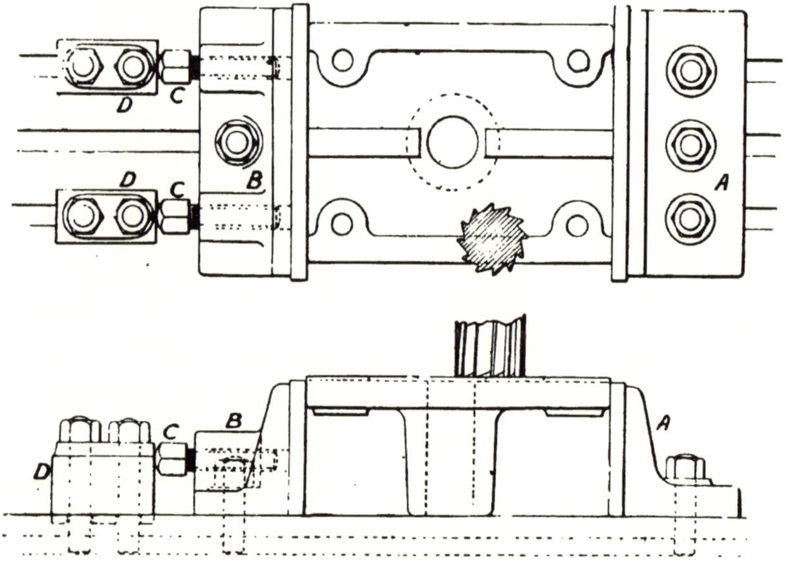

Fig. 196.—End Mill facing a Boss Bearing pinched between Special Clamps, with Stop Blocks.

of the machine, Fig. 195, and bolted down with clips near the ends, the clips clamping adjacent plates as shown. The longitudinal edges already tooled must be set square across the table. Both sets of ends will be milled at the same time. The lengths of these plates must be the same as the lengths of the square plates, and will be measured with a gap gauge of sheet steel.

Fig. 196 shows a boss bearing which cannot be gripped on top nor on edge without special angle brackets. One of these, A, is fixed by tee-head bolts in the table slots; the other, B, is permitted a limited amount of adjustment before it is tightened by the reaction of the screws C, C against the stop blocks D, D bolted in the grooves.

Fig. 197 shows how eccentric straps are machined preparatory to being bored. A number of half-straps are bolted to an angle plate by means of a clip, the joint faces being levelled properly. After the joints are milled, the bolt holes are marked and drilled at the same setting on the angle plate.

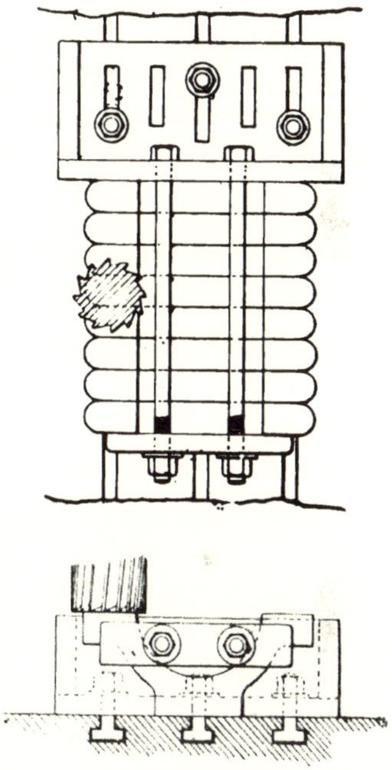

Fig. 197.—End Mill tooling Joints of Eccentric Straps clamped against Face of Angle Plate.

There are very many types of that kind of work which are comprehended under the term "brackets." They constitute a large class. Many bearings are so denominated, rather loosely; but such is the usage. The method of fixing brackets for machining will depend partly on their shape and proportions, partly on the number of similar castings. Small brackets, of which there are but one or two castings, will generally be either bolted to the angle plate or held in the vice. If numerous, they will be arranged in series on the milling machine, and a continuous cut taken over the lot. Shallow brackets will be tooled with the foot uppermost. Deep brackets are tooled while laid upon their sides, the feet standing perpendicularly, and then they may either be milled superimposed, or end to end, or in some cases an alternative will be to mill deep brackets with their feet uppermost, lying horizontally while bolted to an angle plate attached to the table of the machine. The reason why the mode of fixing shown in Figs. 198 and 199 is not suited for deep brackets is that it would not be firm enough to ensure steady cutting.

Fig. 198 illustrates two brackets held down for milling the feet. In this case the brackets have been already bored, which, when practicable, is generally done before tooling the feet. It cannot always be done, but in very many cases—in most instances

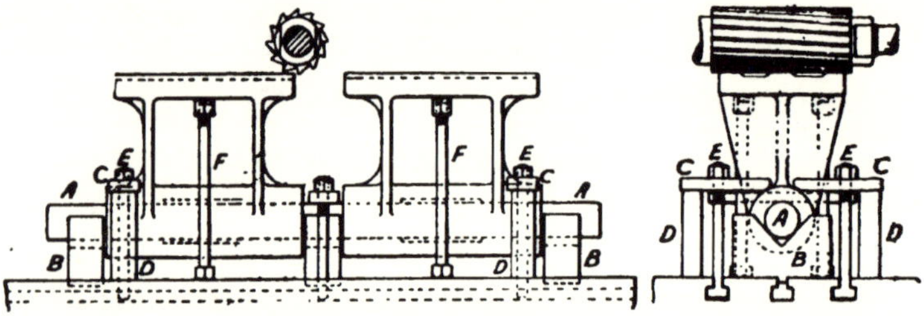

Fig. 198.—Edge Mill tooling Feet of Brackets carried on a Mandrel in Vee Blocks, Mandrel clamped down.

probably—it is as easy to bore first as afterwards. Then, supposing the setting is done accurately, the foot is bound to be parallel with the bore.

Through the bored holes in the brackets a turned shaft A is

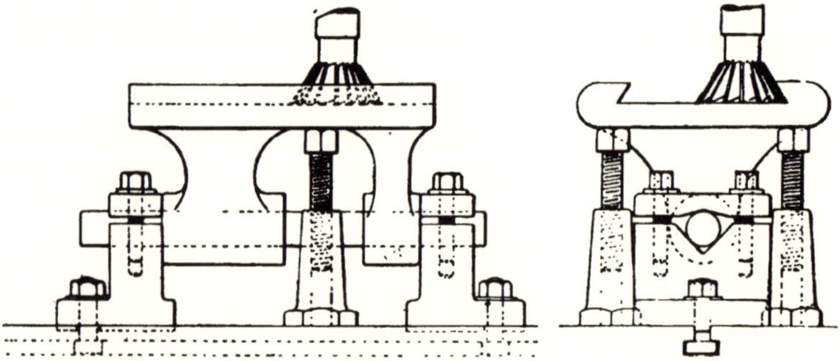

Fig. 199.—Angular Mill facing Bracket supported in Vee Blocks.

thrust. The ends of this are laid on vee blocks B, B, and plates C, C rest upon the bosses and upon packing pieces D, D, and are held down with bolts E, E in the tee slots. In a space left in the middle between the two brackets a clip, laid directly upon the shaft, is pulled down with two bolts.

These suffice to hold the shaft A down, and with it the brackets. But the latter would turn or cant upon the shafts under the stress of cutting, and so it is necessary to pack them at each side under the feet. Pieces of bar might be used, but they must either be of the precise length required, or else the length must be made up with thin iron or steel wedges inserted under their ends, or bolts F may be used, the precise adjustment for length being made by turning the nuts.

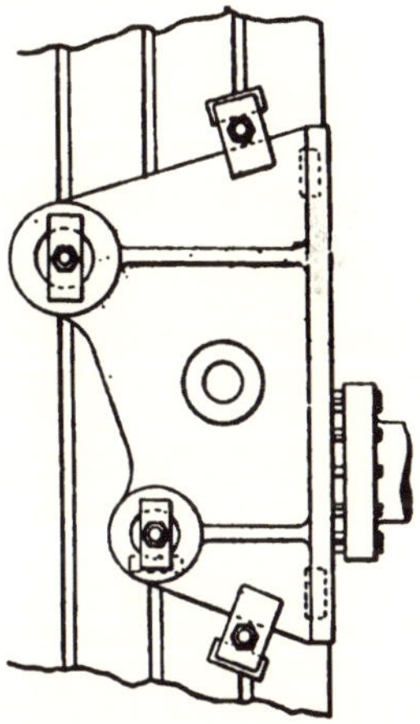

Two brackets only are shown in Fig. 198. More than two can be arranged in line, taking care to have a sufficient number of strips for clamping. Also, two similar rows can be clamped side by side, and two cutters be brought into operation at once.

Cases arise in which it is not practicable to insert a shaft through holes, because the latter cannot be bored until the brackets have been tried in place, in which case the holes will be either rough cored or left solid. Then the brackets can be held with side clips, still being sustained from canting sideways by means of bars or bolts.

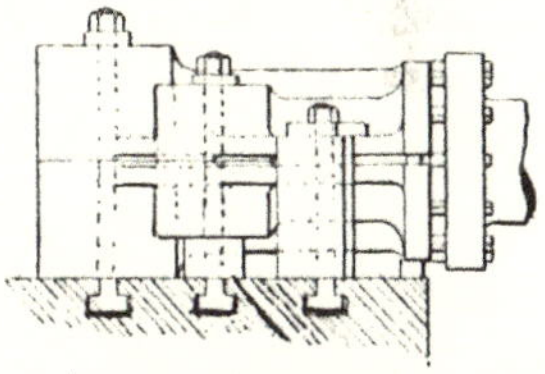

Fig. 200.—Two Brackets, set Back to Back, being faced with a Rotary End Cutter.

Fig. 199 illustrates a regular jig for bracket work. Capped vee blocks are bolted to the table, and the proper adjusting screws take the place of the bolts F in the previous figure.

Fig. 200 is an example of two double bearings set back to back, and bolted and clamped down thus together to the table of the machine. Such bearings could be set end to end simply, and milled in line. This method would be more economical of time than that shown in Fig. 198, because the cutter would only have to travel about half the distance. But the method shown in Fig. 200 has this counterbalancing advantage—the brackets being set back to back, are in the precise position which they will occupy during boring, and also when bolted in their places to receive their shafts. The alignment of the brackets is therefore assured by doing as shown, while adopting

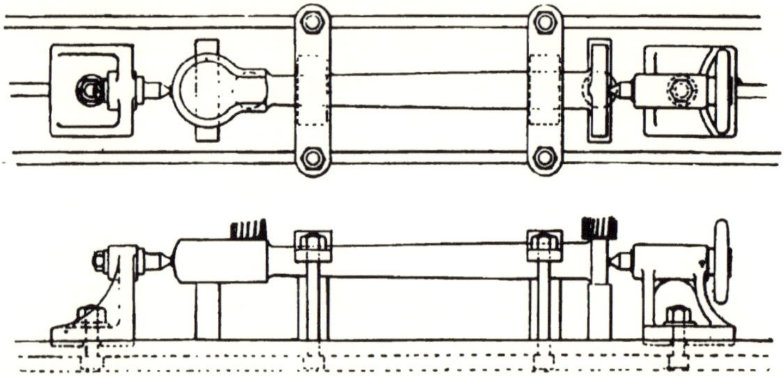

Fig. 201.—Bellied Connecting Rod being milled on Centres.

the end to end method there is just a risk that after the feet have been milled the holes might be bored out of line: or, if set truly for boring, they might not bore out clean, or "hold

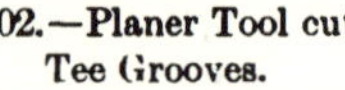

Fig. 202.—Planer Tool cutting Tee Grooves.

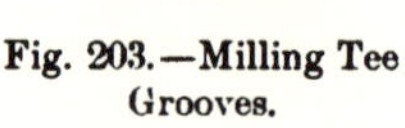

Fig. 203.—Milling Tee Grooves.

up" to dimensions everywhere. Hence the reason for the adoption of this method.

Or again, the brackets might be bolted to an angle plate. But being in line, the same objection would exist.

The joints of fly-wheels, cast in halves, are often milled in similar fashion to Fig. 200.

Fig. 201 illustrates the use of centres in milling the faces of a connecting rod. Vee blocks might be used, if the rod were not bellied. It is centred by the centres left from the lathe, is packed on blocks at each end, and clamped down, blocks being inserted immediately under the bolts.

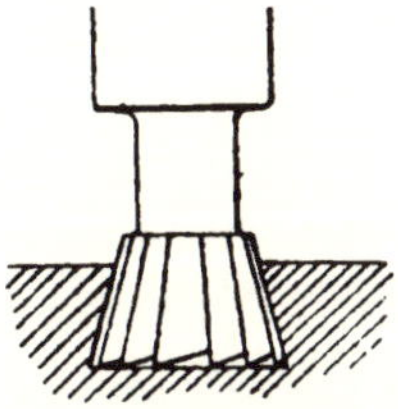

Fig. 204.—Milling Dovetailed Grooves.

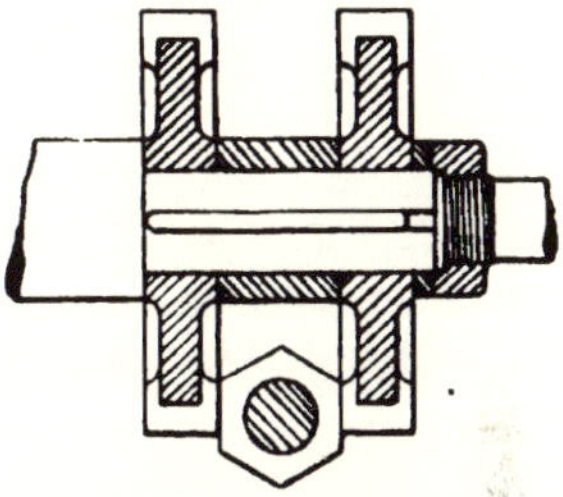

Fig. 205.—Nut Milling with Straddle Mill.

Fig. 202 illustrates the familiar method of planing out tee grooves in machine tables, Fig. 203 shows how they are milled. Fig. 204 is a slot of another kind being similarly treated. In each

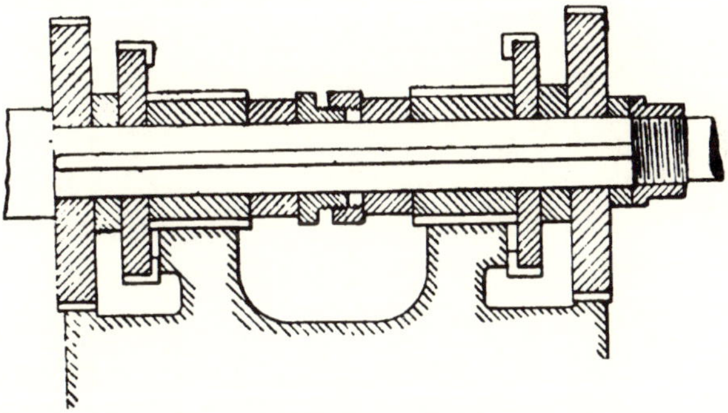

Fig. 206.—Gang of Six Cutters, with Adjusting Screws, operating on Six Faces of a Bed.

case the milling cutter has the advantage of sizing as well as shaping.

Fig. 205 shows a straddle mill cutting nut faces, thus sizing as well as shaping. No lining out is wanted, but a division plate sets the exact angles. Fig. 206 is an example of a built-up straddle mill tooling six faces, and edges of a bed. Adjustment for side wear is by the nuts at the centre of the arbor.

Small work is generally held for tooling in the vice, rather than by bolting by the tee grooves. There are two main classes of vices—those which swivel, and those which do not. The first are employed on the majority of jobs, but the second are often handy. The ordinary milling machine vice is a squarely constructed appliance, the hinder jaw sliding over broad surfaces. There are many special forms. In one of the latter, made by the Garvin Machine Company, the fixed jaw is adjustable on the base by means of tee grooves, to facilitate rapid adjustment. In another the vice is made to swivel, but the swivelling can be

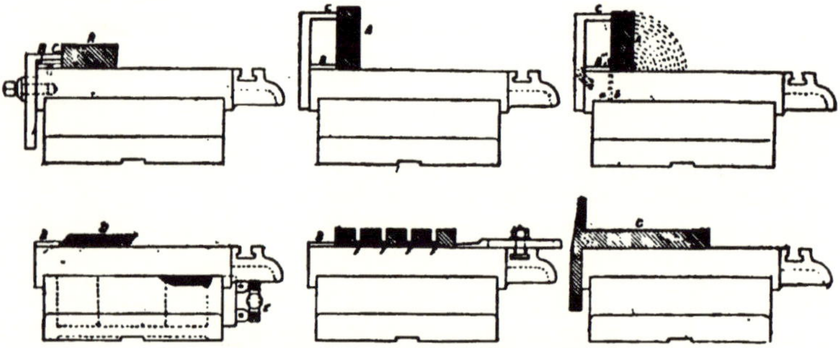

Fig. 207.—Walker Magnetic Chuck.

either in a horizontal or vertical plane, by the simple plan of having two bases at right angles with each other, which brings the plane of the swivelling face into the vertical or horizontal.

In the Brown & Sharpe Company's universal vice the upper part consists of two portions: the lower, which swivels in a horizontal plane; the upper, hinged to the lower to swivel in a vertical plane anywhere between the horizontal and 90°. The clamping of this portion of the vice is done by tightening the hinge bolt, and also by a bolt which connects a double-bracing lever, by which the freely moving part of the vice is connected to the lower portion.

In another type of universal vice, the base, which swivels in a horizontal plane, has two uprights cast upon it, between which the upper portion pivots on a pin running right through. Two quadrant slots in the uprights receive a through bolt, by which the upper part is allowed to swing through 100°, and permits of clamping in any position.

Figs. 207 show end views of the No. 5 standard type magnetic chuck by O. S. Walker & Co. The underlying principle of these chucks is clear from the figures. The first one shows the chuck set up for surfacing the flat side of the parallel piece A, which lies against the back strip B, and the vertically adjustable back rest C, the edges of which are exactly square with the chuck top.

In the next figure is shown the method of holding the piece A, for finishing one of the edges—note that its bottom edge is slightly bevelled. When placed on the chuck, with the acute angle outward, this piece is magnetically held, not only downward against the chuck top, but also against the back strip B, and back rest C, which latter has been elevated to come near the top of the piece. In this manner the top edge of the piece can be tooled square with the side. The magnetic action on the piece when adjusted in this position is illustrated in the next figure. The dotted circles and arrows serve to indicate the course of the magnetic lines of force. The broken lines *a*, *b* represent the positive and negative magnetic poles, the magnetic force seeking an easy path across the gap between these poles. The lines of force therefore preferring a metallic circuit to an air gap, travel through the piece of work A, and then dividing and passing down through C and B to the other pole of the magnet. Owing to the peculiar construction of the magnet face, a strong side pull is obtained as well as a direct down pull of the work. In milling, this piece would require additional staying at the side to hold it in place, and across the end of the chuck there is fastened a vertically adjustable strip that forms an end stop, this not being shown in the drawing.

The next figure illustrates the chuck with the back rest C removed, and the chuck holding the strip D, for finishing one of the flat sides. A portion of the end of the chuck is broken off to show the coil chamber. E represents an ordinary knife-edge switch, which is protected by a guard on the edge of the chuck as shown.

The next figure shows the method employed for holding a number of parallel pieces at one setting. In this case additional holding power is got by laying a strip of non-magnetic metal F between the pieces to prevent the magnetic lines from flowing horizontally through the pieces toward the back strip B, thus diverting the power from the face of the chuck. This view also shows one of the slotted fingers for staying the work and preventing the side slewing. The last figure illustrates how the piece G, with a right-angular projection, can be held.

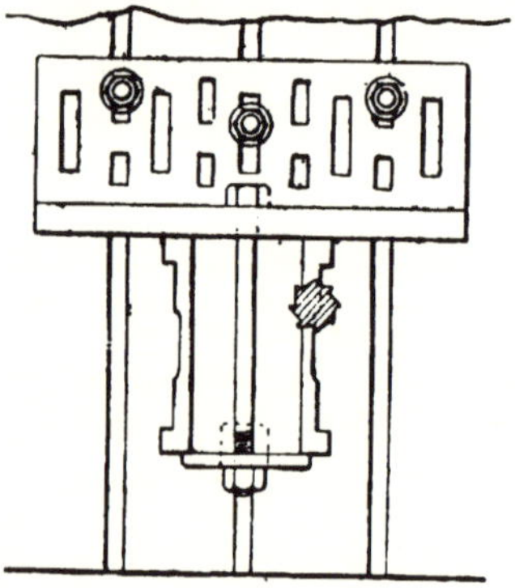

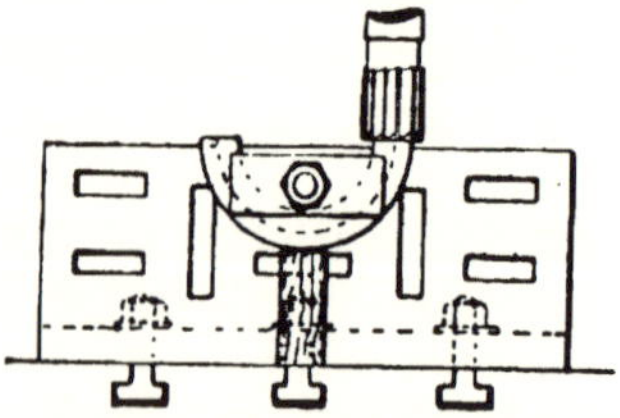

Fig. 208.—Brass held with Angle Plate, for milling Joint Face.

In case the chuck is to be used with water, the coil is given a waterproof treatment and the terminals are capped over entirely, leading the wires through a rubber hose to some convenient position outside of the area of the water where the switch may be attached.

A duplex switch is used with these large chucks. This is very desirable, especially for large work which is difficult to remove from the chuck. It consists merely of a double throw with cross connected wires, so that, when the handle is thrown completely over, the current is led through the coil in the opposite direction. The contact must be timed exactly right to prevent recharging the chuck face with the polarity reversed.

Fig. 208 illustrates the milling of the joint faces of brasses instead of putting them on the shaper. The half brass is clamped against an angle plate, and additional steadiness secured by a piece of wood packing at the front end. In Fig. 209 a cap is clamped through the medium of a clip, and the joint face and edges and cap faces are milled.

So much clutch work is done that the old style of using the teeth as they are cast, without tooling, is not considered good enough for any but the roughest class of work. Even though the teeth are cast, as in the larger clutches, they are tooled subsequently, the case of spiral teeth excepted. At one time all such work was done at the shaper, and is so done still in some firms: the clutch being held in the vice, and turned about to present successive faces to the tool. In the absence of dividing apparatus

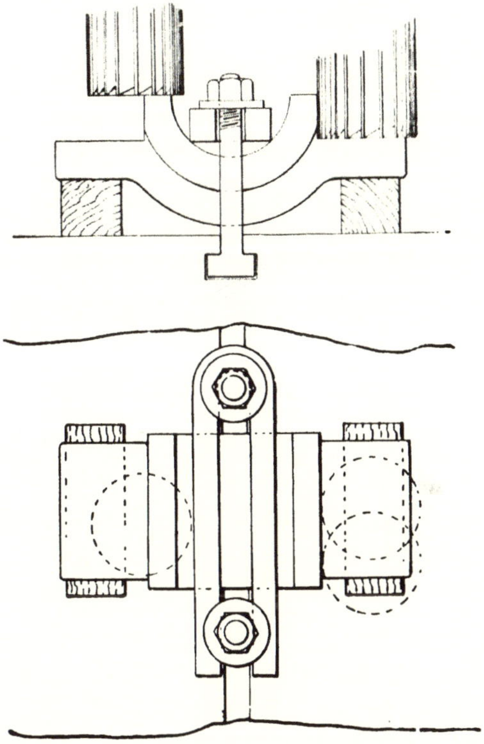

Fig. 209.—Cap held with Clip for milling Faces and Edges.

the jaw edges were lined out, and a templet kept to check them by. Such work is now often properly appropriated by the milling machine, using an indexing head for effecting the divisions correctly, without lining out or templating.

Fig. 210 illustrates one method of milling the teeth, which applies alike to clutches having the teeth cast, and to forged or cast

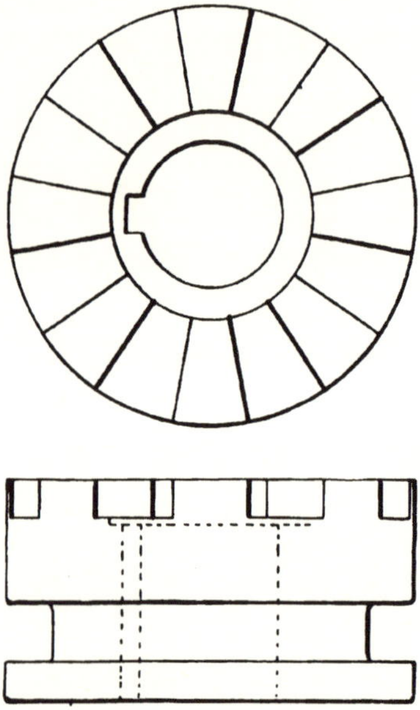

Fig. 210.—Claw Clutch to be Milled.

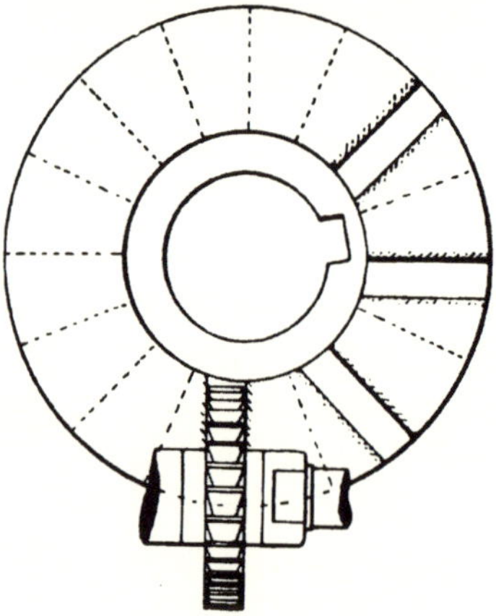

Milling Claw Clutch.

ones having the teeth milled from the solid. The work is clamped as in Fig. 191, page 211.

A cutter is selected having teeth on the edge and sides, its axis being set at right angles with that of the clutch. Its width must be a trifle less than the space between adjacent teeth at the inner part, next the hole, where the space is narrowest. It is then set to mill one side of a tooth, and fed across down to the full depth of tooth, or nearly so, depending on whether a fine finishing cut will be taken all over the bottom subsequently. The table is then run back, and the index set for the next tooth, and so on until the sides of all the teeth which face in that direction are done. The other side of the cutter is

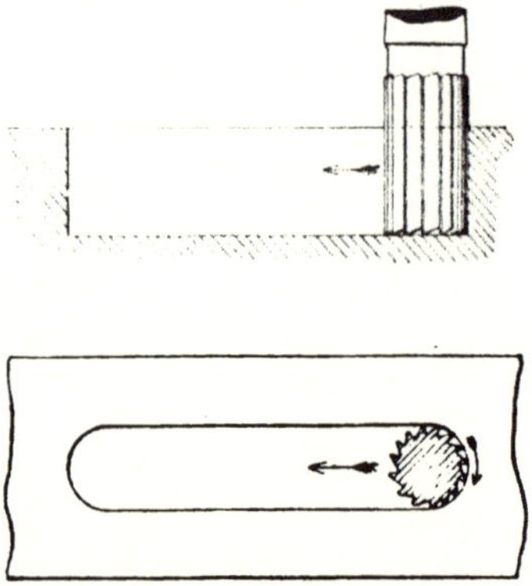

Fig. 211.—Milling a Slot.

next set to do all the opposite sides of the teeth in the same manner. Finally the bottoms are finished with the same cutter.

Fig. 211 is an example of slot milling *versus* slot drilling.

The formation of circular bossed ends is often done on the slotting machine and on the shaper. It is usually more convenient to set small levers on the shaper for bossing, and large ones

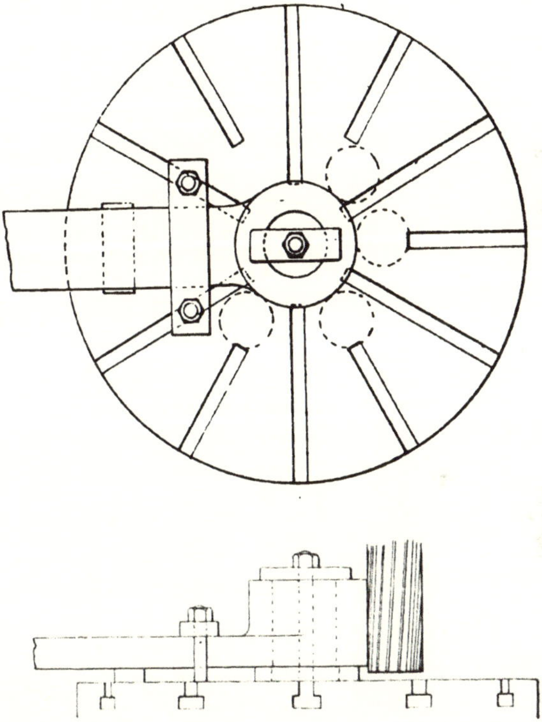

Fig. 212.—Illustrates Value of Circular Table for Boss Milling.

on the slotter. There is rather more difficulty in setting them on the latter than on the cone mandrel of the shaper, which is self-centring. On the slotting table, measurement alone is frequently the only means of setting. Many tables with a central hole are, however, fitted with a central arbor or mandrel, standing up vertically, upon which the bored holes of bossed ends are centred. Such arbors are turned parallel to different diameters to suit different jobs, or one of a constant small diameter can be used, and bushes slid over it for holes of larger diameter.

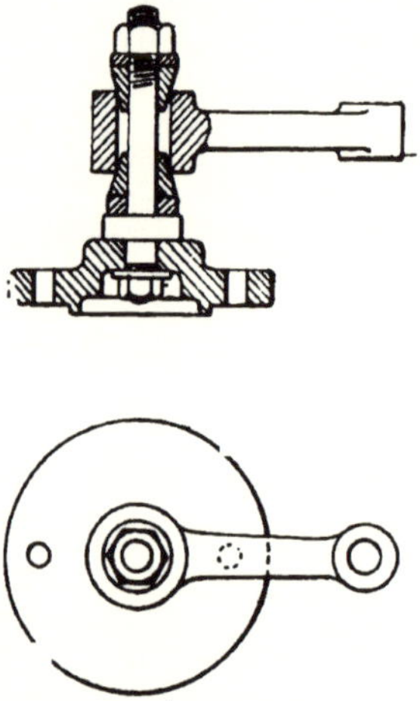

Fig. 213.—Cone Mandrel to bolt on Circular Table.

Fig. 212 shows a common job which is preferably done by milling, that leaves a smoother surface than slotting. When such a piece is slotted, the boss is turned down as far as the web. This is not necessary in milling, for by altering the height of the cutter the boss can be milled all round. Figs. 213-215 illustrate useful attachments for milling circular work, borrowed from the shaper arbor. Fig. 213 is a fixture for bolting directly on the centre of the circular table. The lever shown is held by the conical centres, and clamped, and this partakes of the circular movement of the table. Fig. 214 is a fixture derived from the previous one, but it is bolted to an angle plate, and holds a double-ended lever

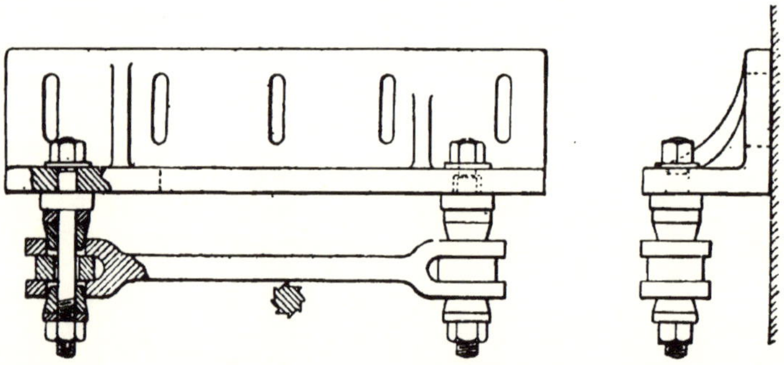

Fig. 214.—Angle Plate with Cone Mandrels for Clamping Levers by for Milling.

while the web is being milled. A washer is inserted between the forks to prevent them from being closed inwards by the bolts.

Fig. 215 is an auxiliary table or plate bolted to the machine table for holding objects having holes, so as to mill both bosses and webs.

The milling machine has invaded the province of the lathe to a slight extent. Some jobs that are not infrequently done are the rims of very light spur-wheel blanks that are rather delicate and springy to operate on in the lathe. Bevel-wheel blanks are also tooled as in Fig. 216. Belt pulleys are sometimes so treated, Fig. 217, and rope pulleys, Fig. 218. In each of these cases the cutters are built up, and the entire rim is tooled at one operation.

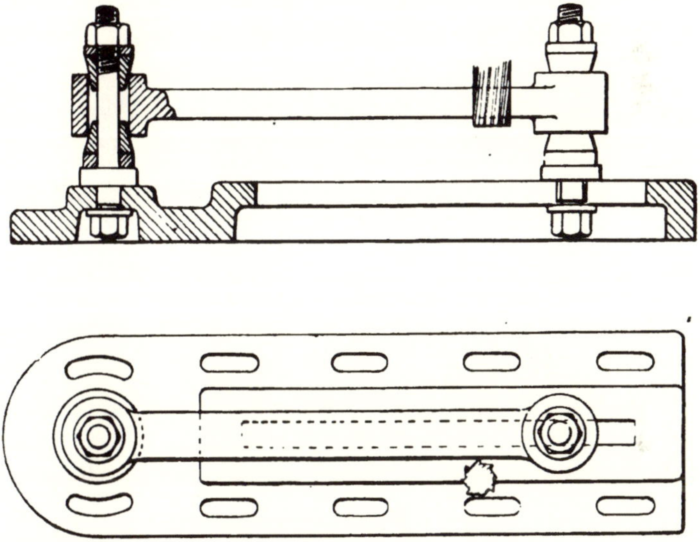

Fig. 215.—Auxiliary Table with Cone Mandrels for Objects having Holes.

Milling is largely taking the place of other machine tool operations in die cutting. Fig. 219 shows a group of cutters used in this work. They comprise tools for tapered edge, and for parallel work, for producing concave portions, for roughing and finishing. They fit directly into the spindle, or into an intermediate fitting shown in the figure.

Fig. 220 shows a job of profiling being done in which the tapered form of tracer or former pin is seen bearing against the form carried on a bracket bolted to the table and controlling the operation of the cutter on a pile of six pieces of work. The tracer pin is adjustable in a slot by the screw indicated. For remarks on profiling machines, see pages 96 and 100 to 104.

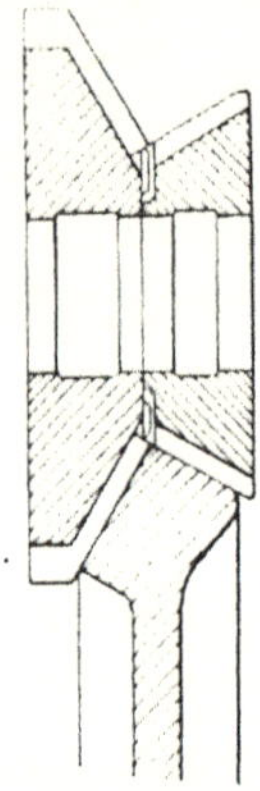

Fig. 216. Milling a Bevel-Wheel Blank.

Holding Work in Jigs.—There are several ways to look at this subject, and many things to be considered in so apparently simple a thing as gripping pieces of work to be milled. Other matters besides holding merely have to be borne in mind. The following are the chief considerations:—

Security of course is necessary; but avoidance of risk of distortion is essential. So is rapidity of gripping. Also a method of holding that will not interfere with the freedom of movement of the cutter or cutters; frequently, too, facility for changing the position of the work to expose fresh surfaces to be milled. The question of adjustment of rough or of tooled surfaces in relation to other surfaces of the work, or of holes, often affects the question. So does that of fixing one or of several pieces for simultaneous milling. Frequently the table of the machine receives the piece or pieces of work, as in the examples previously given. But in many instances this would not afford sufficient conveniences for holding, as in cases where numerous important and necessary relations of parts have to be taken account of, and then the special jig has to be devised for one particular kind of article made in quantity, or sometimes for articles that are not very dissimilar in shape or size.

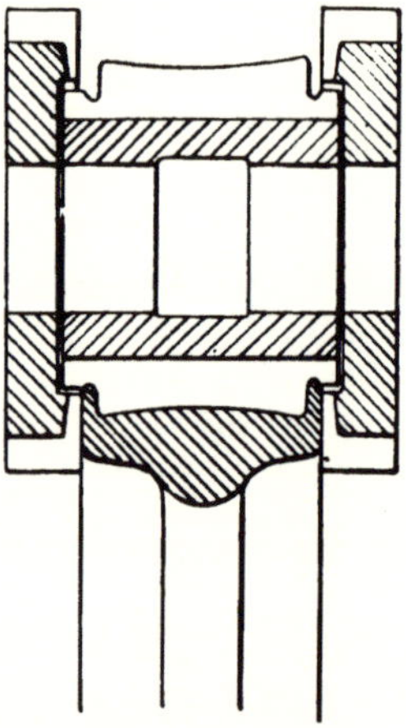

Fig. 217.—Milling the Rim of a Belt Pulley.

Going a step further, the same jig is often designed to include another operation besides milling—that of drilling of holes, effected at the same or another setting in the jig, which is done in order to ensure the correct relations of such parts without resorting to measurement or check.

Then further, in another stage, a gauge piece is sometimes included, by which to adjust the relations of the mill and the work, and so to ensure correct height or thickness for tooling, without resorting to direct measurement.

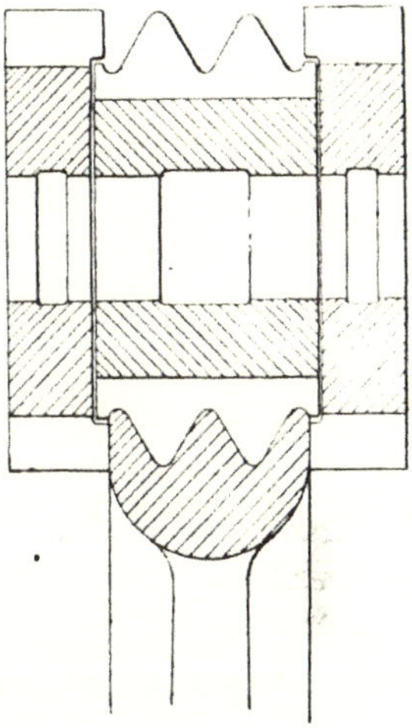

Fig. 218.—Milling the Rim of a Rope Pulley.

The first and most obvious method of holding work directly on the tables of milling machines, with only the aid afforded by the tee grooves, has been already illustrated. The clamping plates or pieces used are held with tee-headed bolts in these grooves, and the clamps may bear on top, or against the edges of the work, either by the sides or the ends. Or the work may be held against the face of an angle bracket or plate clamped to the table. Pieces to be milled circularly are held by an arbor in the centre of a rotary table. Holes are frequently drilled first to receive plain or conical arbors inserted in the circular table, or attached to angle plates. If jigs are used, these are provided with a tongue in the bottom to fit the table grooves, to which they are bolted down. These summarise the general methods of holding work for milling.

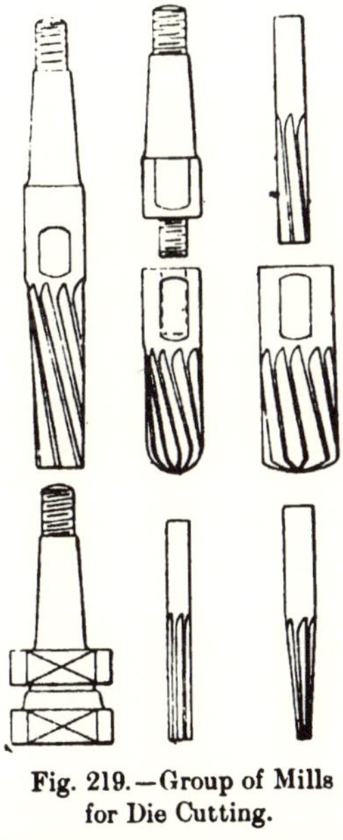

Fig. 219.—Group of Mills for Die Cutting.

It is more difficult, as a rule, to make suitable jigs for castings than for rolled metals, as shafts and plates, or for drop forgings. This is due to the fact that variations occur in nearly every dimension of the separate castings, though made off the same pattern. From experience I should say that no two castings, when hand made, are ever exactly alike when brought to the test of the jig maker. With good machine moulding they are generally uniform, though not always. Variations increase with increase in dimensions. This difference in hand and machine work is that due to rapping in delivery, and mending up in the first case, and the coercion exercised by the machine lift in the second, and the fact that mending up is rarely attempted. The reason why slight differences

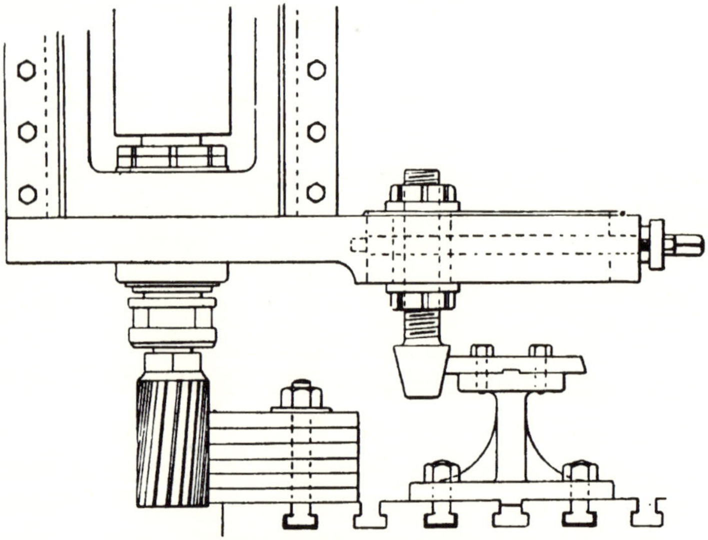

Fig. 220.—Milling with a Former.

do sometimes occur in machine-made castings is that differences in metal involve variable shrinkage amounts, and changes occur in the mould in the act of pouring, such as scabbing, swelling, producing rough surfaces; and lumps, straining of boxes increase the dimensions. Increase in dimensions of castings gives larger masses on which these causes operate.

The jig maker has to be on his guard, therefore, when making provision for holding castings, and his appliances must embody some elasticity in design to permit of the embracing of castings that vary in dimensions, even if only by slight differences.

The methods generally adopted include vee'd spaces for bosses, and adjustment screws. Bosses generally when present are taken as the parts from which to commence the settings, in order to locate their holes centrally. It looks bad to see holes out of centre, whereas a trifle of difference in the thickness of a foot or facing is of little moment.

Fig. 221.—Jig for Locking-Plate Castings.

In some jobs that are nominally left rough cast on a lower face it is often advisable to take a light cut over that surface, simply for convenience of clamping the face down to the table, to avoid packing up. The same may often be done with advantage when the face has to be finish-tooled subsequently. This applies more particularly to light flimsy castings, or forgings which are most liable to spring. Such precautions are more necessary in the case of milled work than in that of work to be tooled on planer or shaper, because of the greater pressure of the cut in milling, with increased risk of distortion. If only the high parts are removed, bedding can take place there, and the clamps be brought opposite them. In some cases this will be quicker, and almost always safer than attempting to pack up with wedges, or thickness strips. For a

Fig. 222.—Turret Fixture.

similar reason it is often found necessary to touch off the high spots or lumps on castings and forgings before they can be inserted in their jigs, even though those localities have to be subsequently tooled all over to dimensions.

Fig. 221 illustrates a very interesting piece of work being done on one of the Herbert horizontal spindle machines, consisting of locking-plate castings for railway signal interlocking apparatus. These have to be interchangeable, and the accuracy of the gang of cutters has to be maintained permanently. The castings are rough on the back and ends, but they are held in a special jig which allows of variations in the rough backs of the castings. The casting is gripped at the two ends, and is supported

Fig. 223.—Turret Fixture.

about the centre against the stress of cutting by spring plungers in the bottom of the jig, which bear against the under surface of the casting, and are afterwards locked by the square-head screws which are seen at the front of the fixture. The cross slots are milled subsequently in the same machine, in another fixture. The interlocking of the cutters on the "hit and miss" style will be noticed, which provides for adjustment to keep the width of grooves constant, to compensate for resharpenings. The stiffness of the outer bracings may be noted, as important elements in the work of heavy cutting.

Figs. 222 and 223 show an advanced piece of milling practice as done in the shops of A. Herbert Ltd. It comprises a turret

Fig. 224.—Fixture for Toggle Levers.

fitting to the table of a horizontal machine, rigged up specially for milling the heads of bolts. The two illustrations show the fitting from two points of view.

There are six cutters—two outside, and two pairs in intermediate positions. The bolts are fastened in the turret, and the latter rotated into six equidistant angular positions, in each of which three bolts have two opposite faces milled simultaneously, the equivalent of a single bolt finished at each position of the turret.

The cover for the cutters includes an arrangement for lubricating them. The finished bolts are removed, and fresh ones inserted while the machine is in operation. One man operates two machines in the Herbert works, but this could be exceeded in a screw factory, or on some classes of work.

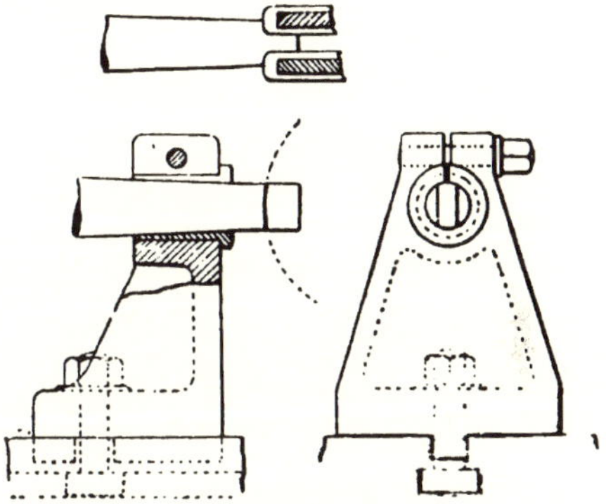

Fig. 225.—Jig for Milling Arbor Tangs.

Fig. 224 illustrates the milling of malleable-iron toggle levers that form a portion of the automatic chuck on the Herbert hexagon turret lathe. Three grooves are milled at once, to fine limit gauges, the work being interchangeable. The cutters are of the inserted tooth kind, alternate teeth cut on opposite ends, thus permitting adjustments to be made to compensate for wear—effected by driving the teeth farther through the body of the cutter in both directions. The clamping of the pieces in the fixture is interesting. The levers to be milled drop into recesses in the jig, and as clamping by the top edges would interfere with the cutter arbor, they are clamped on rods coming out at the sides. The rods are inserted in holes previously bored in the pieces.

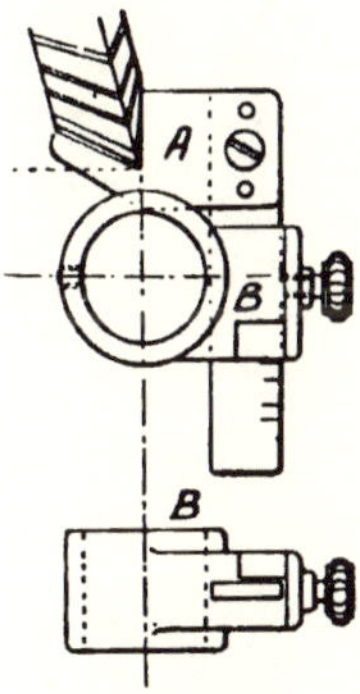

Fig. 226.
Setting Gauge.

A neat rig-up for milling the tangs on the ends of taper shank drills, arbors, reamers, and which is adaptable alike to the square necks of hand taps and hand reamers, is illustrated in Fig. 225. It consists of a bracket, fitting on the table, with a tongue for the table grooves, being thus set in line with the spindle hole in the index head, and clamped in position. It is bored to receive

Fig. 227.—Fine Gang Milling.

a split bush which takes the end of the drill, or reamer, &c., and which is clamped by a set screw in a split lug. The cutters are

Fig. 228.—Jig for Profile Milling.

indicated in their relation to the work. Bushes of various sizes can be substituted, all fitting alike in the bracket, but having holes of different sizes, as many as may be required.

Fig. 226 illustrates a jig in the Ludwig Loewe shops for setting cutters used for making the teeth in spiral mills. It comprises a graduated beam with a fixed jaw A, and a movable jaw B. The graduations on the beam give the diameter of the cutters to be

Fig. 229.—Link Type of Jig.

milled, so that, when the movable jaw is clamped to a given graduation, one element is fixed. The other—the position of the

Fig. 230.—Milling Headstock Foot with a Gang of Cutters.

cutter—is determined by the notch in the fixed jaw A, and when the machine is set by these two positions, the work can be commenced; and continued without any experimental cuts being required. The hole in the sliding jaw takes bushings that fit the arbors on which the cutters are placed to be milled.

A good many devices have been schemed to permit of a milling machine attendant fixing work on one end of a jig, while the machine is milling similar pieces at the end opposite; an idea which is sometimes utilised on double-headed planers. The advantage lies chiefly in forgings and castings, which, being more or less rough and uneven, take more time in their adjustments than pieces of sheet metal, or pieces partly machined would do.

An example of extremely fine work in gang milling is seen in Fig. 227, done in the shops of A. Herbert Ltd. It is the cutting of slots in burners for gas fires. Large numbers of slots were cut in one traverse, the widest burner having 120 slots, requiring the fitting of 120 saws on one arbor. The difference in time between this and cutting single slots is enormous.

The photo in Fig. 228 illustrates profile milling on a Herbert machine, the work consisting of shaping the edges of the links for chain conveyors, two of which are seen lying in front of the fixture, which is a vice especially designed for holding them. The links are first drilled uniformly in a jig, and they are then located in line by bars passing through the holes. The cutter is made in three parts. As the work is broad, the necessity for a very rigid support to the cutter is obvious. How this is ensured is seen in the illustration, the bracing affording a support as rigid as a solid housing could be.

A very interesting photo is that shown by Fig. 229 from the practice of the same firm. It is a special fixture made for the horizontal milling of objects of irregular outline, the particular example on the machine being a mould such as is used for pressing tiles. It is concave in form, the radius being very large. The apparatus is mounted on links resembling those of an ordinary ruler, thus ensuring a parallel motion. On the under side of the table on which the work is mounted there is a former, the edge of which is seen, and its face is the exact complement of the object being milled. The former can thus be made in the first instance, if convenient, by fitting it into an actual piece of work. It rests upon a roller which is carried upon the knee of the machine, the outer end of which is visible. As the machine table is fed transversely carrying the jig, the former in passing over the roller gives the required curvilinear movement to the face of the work, and the parallel links accommodate themselves thereto, while coercing the work rigidly in other directions.

Fig. 231.—Milling Quadrant Piece for Change Gears.

There are other interesting points to note about this job. One is the clamping down of the plate. Two of the clamps are seen in the front, and it will be noticed that their lower edges work on an incline, causing them to pull the work down when the nuts are tightened up.

The very coarse pitch of the cutter is remarkable. This subject has been discussed on page 164, and is a feature which is likely to grow in favour. Messrs Herbert say that in their experience very much coarser pitch cutters can be used with advantage than is common practice.

Fig. 230 illustrates a fixture for milling the feet of headstocks with a gang of cutters in the Ludwig Loewe Works. Fig. 231 shows the tooling of both faces of a quadrant piece with built-up cutters in the Cincinnati shops, the circular table being used to rotate the work.

CHAPTER IX.

INDEXING, SPIRAL WORK, AND WORM, SPUR, AND BEVEL GEARS, ETC.

The Universal the Machine of Applied Geometry—The Spiral Dividing Head—The Basis of Calculation—Worm and Worm Wheel—Index Plate-Sector—Differential Indexing—Angles of Spirals—Rules and Examples—Graphic Method—Milling Screw Gears—Relations between Spiral, Helical, and Worm Gears—Elements of the Spiral Gear—Velocity Ratio—Pitches—Graphic Methods — Examples — Trigonometrical Rules — Worm Gears—Methods of Cutting.

The Machine of Applied Geometry.—The milling machine has been termed "the machine of applied geometry," because of the combinations which are possible in a universal. The universal is a notable exception to the general rule that machines that combine many functions are not so economical as those which deal with special operations. There is hardly any operation of the machine shop which is not done on the universals by the aid of the numerous adjuncts that are legitimately fitted to it, that is, they are not of the nature of makeshifts. It includes plane surfaces, and circular, both external and internal, drilling, reamering and boring, thread cutting, key grooving, profiling, gear cutting of all classes, and such work is done at all angles. It includes the correct spacing of holes and distances, and much besides, regularly done on pieces of medium dimensions.

The advantages are most apparent in work requiring several settings or spacings, as an alternative to lining out by hand methods. The latter are never strictly accurate, and are generally tedious. The milling machine is both accurate and rapid.

The Spiral Dividing Head.—The leading elements of the spiral dividing head are the rotation of the spindle irrespective of the angle at which it may be set, the change gears by which a precise and definite rotary motion may be imparted to the spindle relatively to the pitch of the lead screw, and therefore to the travel of the table, and the divisions effected by the index plate.

The Basis of Calculation.—The basis of calculation is the single-threaded worm, and the 40-toothed worm wheel, which by general consent, and its great convenience in subdividing into thousands, is adopted in milling machines. As the worm thread and the worm-wheel teeth are as 40 to 1, forty turns of the crank produce one revolution of the spindle, and one turn of the crank a fortieth of a revolution of the spindle. To obtain forty divisions to the circle, therefore, the handle would be turned forty times. To obtain eighty divisions the crank would be turned half-way round for each division, but, to obtain twenty, two turns would be required. The rule is therefore:—

Divide 40 by the number of divisions required, and the quotient will be the number of turns or parts of a turn which will have to be imparted to the crank. Thus—

$$\frac{\text{Wheel, 40}}{\text{Number of divisions, 40}} = 1 \text{ turn.}$$

$$\frac{\text{Wheel, 40}}{\text{Number of divisions, 80}} = 0{\cdot}5, \text{ or half a turn.}$$

$$\frac{\text{Wheel, 40}}{\text{Number of divisions, 20}} = 2 \text{ turns.}$$

By means of the index plate a turn of the worm shaft can be divided into a very large number of equal parts, and the fortieth of a revolution of the spiral spindle correspondingly subdivided. Index plates furnish several circles of holes, from one of which selection can be made for the majority of jobs required, that will permit of obtaining equal divisions. Thus, for eighty divisions, the crank being turned half-way round, a circle will be selected that will divide into two equal parts. If three divisions, a circle must have a number of holes divisible by three, and so on.

To facilitate counting the divisions in the circles of holes is the object of the sector A, Fig. 232. Circles of holes on the index plate being selected which will divide equally as even numbers, or multiples of even numbers, the sector is used to avoid the counting of every hole at each partial rotation of the crank.

If 144 divisions are required, then $\frac{40}{144} = \frac{5}{18}$, and thus the index crank has to be moved $\frac{5}{18}$ of a turn to obtain each of the 144 divisions. A circle of holes is selected containing either 18 holes or a multiple of 18, and 5 spaces are measured off on an 18-hole circle, or 10 spaces on a circle of 36 holes.

Or, to cut 96 teeth—

$$\frac{40}{96} = \frac{10}{24}$$

The index pin must be moved $\frac{10}{24}$ of a turn, or, selecting a 24-hole circle, 10 holes for each division.

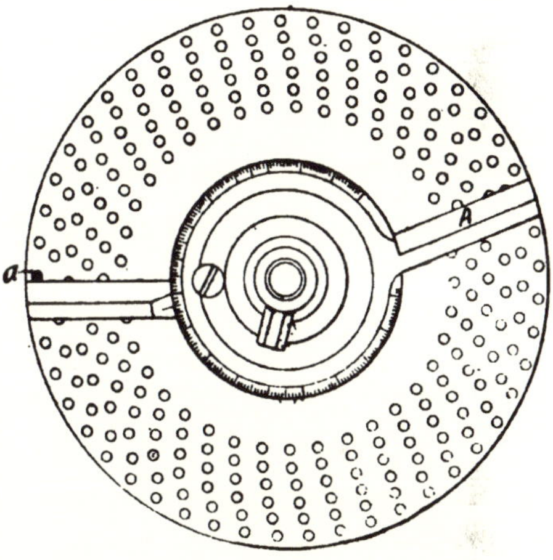

Fig. 232.—Index Plate with Graduations.

Instead of counting every division of five or ten, as the case may be, the sector is set once, and pinched by its set screw, and then afterwards it is simply moved round into contact with the index pin, and then, after a cut has been taken, it is moved round and the pin inserted in the next division for the next cut. Only in setting the sector first, one more hole than the number required for the division must be taken, because that has to be occupied by the pin (see Fig. 232, *a*). The operation of turning and setting has to be repeated as many times as there are divisions required—144 times in the first example, 96 times in the second given, and in each case the spindle and the work will have made a complete revolution. This is the method of simple or plain indexing.

The index tables are furnished with index plates, and save calculation for a large number of divisions. But they do not cover exceptional cases, nor, in fact, all the divisions which are possible for any one circle of holes. This matter is discussed at length on page 248.

The tables supplied by makers of milling machines give but one circle of holes for any given division. But it is not necessary to change a plate simply to place on the one named in the tables. A simple calculation will show whether divisions can be effected with the plate that happens to be on the machine. Thus there are six circles on the B. & S. machine which are divisible by 3, namely, 39, 33, 27, 21, 18, and 15, though the index table gives the first only. Mr E. Gauthier prepared a table to save the slight trouble of calculating, which, with some improvements made by the *American Machinist*, is given below.

Division.	Circle.	Turns.	Holes.	Division.	Circle.	Turns.	Holes.	Division.	Circle.	Turns.	Holes.
2	any	20	...	12	15	3	5	25	20	1	12
3	39	13	13	13	39	3	3	26	39	1	21
	33	13	11	14	49	2	42	27	27	1	13
	27	13	9		21	2	18	28	49	1	21
	21	13	7	15	39	2	26		21	1	9
	18	13	6		33	2	22	29	29	1	11
	15	13	5		27	2	18	30	39	1	13
4	any	10	...		21	2	14		33	1	11
5	any	8	...		18	2	12		27	1	9
6	39	6	26		15	2	10		21	1	7
	33	6	22	16	20	2	10		18	1	6
	27	6	18		18	2	9		15	1	5
	21	6	14		16	2	8	31	31	1	9
	18	6	12	17	17	2	6	32	20	1	5
	15	6	10	18	27	2	6		16	1	4
7	49	5	35		18	2	4	33	33	1	7
	21	5	15	19	19	2	2	34	17	1	3
8	any	5	...	20	any	2	...	35	49	1	7
9	27	4	12	21	21	1	19		21	1	3
	18	4	8	22	33	1	27	36	27	1	3
10	any	4	...	23	23	1	17		18	1	2
11	33	3	21	24	39	1	26	37	37	1	3
12	39	3	13		33	1	22	38	19	1	1
	33	3	11		27	1	18	39	39	1	1
	29	3	9		21	1	14	40	any	1	...
	21	3	7		18	1	12		Degrees.		
	18	3	6		15	1	10	1	18	...	2

An improved form of division plate has been designed by Messrs Brown & Sharpe for attachment to the machine, the feed and speed mechanisms of which were described on pages 34 and 35. One object sought is to avoid the difficulty which sometimes occurs when a cut has to be started at a definite point of the work and the index pin will not drop into a hole. To obviate this, the index pin is made adjustable by means of the screws A, A, Fig. 233.

Another improvement does away with the necessity for counting the holes when setting the sector, a process just described, page 245. A graduated circle is fitted, Fig. 232, and a column is provided in the index table for the machine, which gives the number of divisions of the graduated circle that corresponds with the adjustment of the sector to span the number of holes required.

Though the graduations do not in all cases bring the sector exactly to the holes required, they are sufficiently close to prevent risk of error in locating the holes.

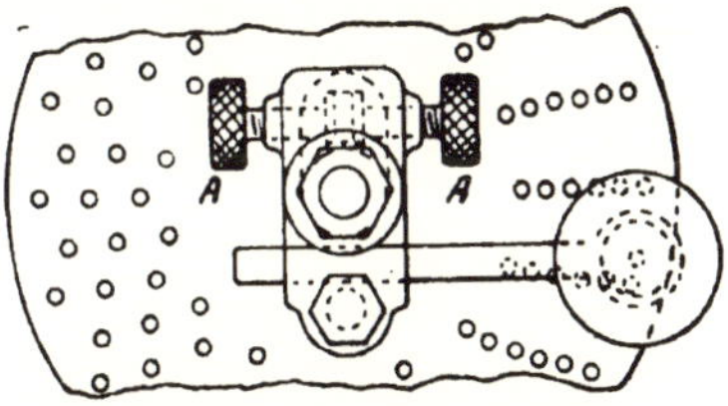

Fig. 233.—Adjustable Index Pin.

The Compound System of Indexing is sometimes used when it is desired to obtain divisions for the work other than those which can be obtained by the direct indexing methods just described, from the index plates furnished with the machine. Compound indexing means that, instead of moving the pin straight ahead on a single circle, two index settings are taken, which involve the addition or subtraction of fractions. Thus the work may be moved through a certain number of spaces in one circle and then turned in the opposite direction a certain number of spaces in another circle. These are indicated in tables supplied with universal milling machines by the signs + and −, the plus sign showing that the two indexings are added together or that the movement of the work in both indexings is in the same direction, while the minus sign shows that we take the difference between the two indexings or that we move one indexing in one direction and the second indexing in the opposite, the principle in making the calculations being simply the addition or subtraction of fractions

The denominators of the fractions indicate which circles of holes are used in the index plates. The following is from Brown & Sharpe:—

"To illustrate the manner of using the machine in compound indexing it may be supposed that we desire to divide the work into 69 parts. Reference to the table (supplied by the firm) shows that the work is moved through 21 spaces or holes in the 23-hole circle, and then turned in the opposite direction 11 holes in the 33-hole circle of one of the index plates. The first movement is made in the ordinary manner. The stop or back pin is placed in the 33-hole circle, the index crank pin is pulled out of the 23-hole circle and the index crank is turned through 21 holes in the desired direction, the holes being measured by the sector. For the second movement, the index crank pin is left in the 23-hole circle, the back pin is pulled back from the plate, and, as the minus sign is given in the table, the crank is turned 11 holes in the direction opposite to that of the former movement. In this part of the indexing the index plate and crank turn together, and, as there is no sector on the back of the plate, the holes or spaces have to be counted directly in the plate. Had the plus sign been given, as in the indexing to obtain 77 divisions of the work, both movements of the crank would have been in the same direction. Ordinarily the order of the movements is not material, and if more convenient for any reason, the back pin could usually be withdrawn first, and the movement described as the second could be made first. In some instances, indeed, for example, in dividing the work into 174, 272, or 273 parts, the outer circle is naturally used first."

Differential Indexing.—A new method of indexing has been recently applied to the Brown & Sharpe entire line of universal milling machines. It is much simpler than the compound method heretofore employed for obtaining divisions that are prime numbers. The indexing is obtained in the same manner as for plain indexing, excepting that the spiral head spindle is geared to the index plate.

Differential indexing differs from the compound method in that the movement of the spiral head spindle in relation to the index crank is positively made by gearing. The index plate is geared to the spindle, thus giving a differential motion that allows the indexing to be made with one circle of holes and the index crank to be

INDEX TABLE—PLAIN AND DIFFERENTIAL INDEXING.

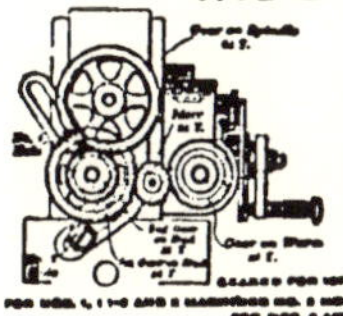

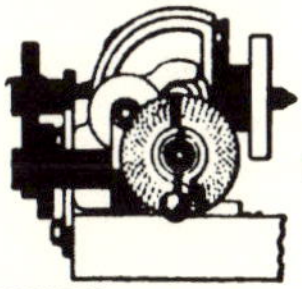

PATENTED NOVEMBER 12, 1901.

FOR USE WITH

UNIVERSAL MILLING MACHINES

BROWN & SHARPE MFG. CO.,

PROVIDENCE, R. I. U. S. A.

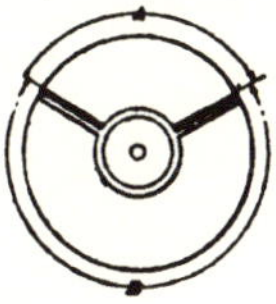

NOTE.

turned in one direction, the same as in plain indexing. This enables spacings to be made that cannot be obtained with an index plate locked with a stop pin in the usual way. When geared for differential indexing, the machine cannot be used to cut spirals, as the spiral head spindle is then geared to the lead screw.

In the differential method there is little opportunity for error. The index crank being moved the same as for plain indexing, it is necessary only to place the proper gears in position, as indicated by the table that accompanies the machine. The change gears and index plates furnished provide for all divisions from 1 to 382, including all the prime numbers, and enable the divisions to be made with no more care than is required in plain indexing. With additional change gears and index plates a large number of divisions beyond 382 can be obtained. In the compound method, however, it is necessary to exercise great care, as two circles are employed, and the direction in which the crank is to be moved, right or left, must be kept constantly in mind.

The Table adjacent is plainly and conveniently made up to include both plain and differential indexing, thus avoiding the necessity of two separate tables. It gives all divisions from 1 to 382.

The index spacing number is 40. In other words, forty revolutions of the index crank are required to make one complete revolution of the index spindle. Therefore, if the index plate is geared to the spindle, using one idler to rotate one turn in the same direction as the crank, and the crank pin enters the same index plate hole, the result will be the spacing number 39, for the reason that while the crank has made forty turns and the plate one, in the same direction, the crank has passed a given point only thirty-nine times. With this same gearing, and the addition of another idler, the motion of the index plate is in the opposite direction to that of the crank, and the plate gains one revolution while the crank has made forty, resulting in the spacing number 41.

With the spacing number 39, it is possible to obtain divisions equal to 3×39 with a circle of three holes, 4×39 with a circle of four holes, &c. This will apply equally well to the spacing number 41. Any division not obtainable with the index plate can therefore be made up with proper gearing.

In general, if the plate rotates in the same direction as the crank, subtract the turns of the plate to one turn of the spindle from the turns of the crank to one turn of the spindle, and the remainder is the spacing number. If the plate rotates in the opposite direction to the crank, the spacing number will be the sum of the turns of the plate to one turn of the spindle, added to the turns of the crank to one turn of the spindle.

Fractional Spacing.—This class of spacing is often required, and, to illustrate the application of the differential method, assume that it is required to space a vernier to read $\frac{1}{12}$ of a degree, or 5 minutes.

Usually the method adopted is to have the vernier spaces occupy a distance corresponding to 11 degrees, and divide this into 12 parts or spaces, which may be expressed as follows: $\frac{11}{360} \div 12 = \frac{11}{4320}$, which is equal to one space. Therefore, there would be $\frac{4320}{11}$ spaces in the whole circle, or $392\frac{8}{11}$ divisions. The indexing of $\frac{1}{9}$ or 360 gives the difference between $392\frac{8}{11}$ and 360, which equals $32\frac{8}{11}$ to be obtained with the gearing. $32\frac{8}{11} = \frac{360}{11}$, which when multiplied by $\frac{1}{9}$, the value of one indexing gives $\frac{360}{11} \times \frac{1}{9} = \frac{40}{11}$ and the proper gears will be $\frac{200}{55}$ or $\frac{64 \times 100}{40 \times 44}$, or gear on spindle 64 teeth; first gear on stud 100; second gear on stud 40; gear on worm 44.

Angles of Spirals.—We have now to consider the method of calculating the angles of spirals for milling cutters, helical teeth, twist drills, reamers, &c., in which the change gears and table screw come in. This mechanism is distinct from that of the spiral head, which effects divisions, or spacing only.

The feed screw of the table has four threads per inch, and must therefore make four revolutions to move the table along one inch. As forty revolutions of the worm in the spiral head are required to give one turn to the worm wheel, and the spiral head spindle; then, if change gears of equal size are used, the table must move lengthwise 10 inches for each turn of the worm wheel and spindle. This relation between a revolution of the spindle and the movement of the table is termed the lead of the screw, or a 10-inch lead; and being constant, is often conveniently adopted as a basis for calculations. In other words, instead of the pitch of the screw, the distance of the table travel for one revolution of the spiral head is employed. Hence the common term, lead of a screw thread, or spiral, equivalent to its pitch, is referred to lead of the machine table.

The change gears are calculated similarly to those of a screw-cutting lathe, the ratio of the driven to the driving gears equalling the ratio of the lead of the spiral required to the lead of the machine. Put in the form of an equation, it stands—

$$(a) \qquad \frac{\text{Driven gears}}{\text{Driving gears}} = \frac{\text{lead of spiral required}}{\text{lead of machine}}$$

or,

$$(b) \qquad \frac{\text{Product of driven gears}}{\text{Product of driving gears}} = \frac{\text{lead of required spiral}}{10}$$

That is, the compound ratio of the driven to the driving gears is represented by a fraction, the numerator of which is the lead to be cut, and the denominator of which is 10.

Or the ratio is as the required lead is to 10.

Or the ratio is as one-tenth of the required lead is to 1.

Or, lastly—

$$(c) \qquad \frac{\text{Driven gears}}{\text{Driving gears}} = \frac{\text{4 times the lead of spiral required}}{40}$$

As in screw-cutting gears, ratios can be broken up into fractions, and numerator and denominator multiplied by numbers to find suitable wheels.

Taking examples in each case and taking equation (*a*)—

Say the lead of the spiral required is 24 inches, then 10 to 24 is the ratio of the gears. Both terms of the factor are multiplied by a number that will give numbers corresponding with the teeth of the change wheels. Thus—

$$\frac{24}{10} = \frac{3}{2} \times \frac{8}{5}$$

Then—

$$\frac{3}{2} \times 24 = \frac{72}{48}, \text{ and } \frac{8}{5} \times 8 = \frac{64}{40}$$

And 72 and 64 are the driven, and 48 and 40 are the drivers; and either 72 or 64 can go on the worm, and either 48 or 40 on the screw, the others going on the stud in the same way that transpositions are effected on the swing plate of the lathe.

Taking (*b*), the rule becomes:—Divide ten times the product of the driven gears by the product of the drivers, and the quotient is the lead of the resulting spiral in inches to one turn. This rule is of value in ascertaining what spiral might be cut with gears at hand. Thus, what spiral would be cut by gears of 32 and 56 driven, and 24 and 40 drivers? Then—

$$\frac{10 \times 32 \times 56}{28 \times 40} = 16 \text{ inches to 1 turn.}$$

Taking (*c*), to cut a spiral with a lead of 21 inches, multiply 21 by 4 = 84. The lead screw must then make 84 turns to 40 turns of the worm. The ratio then stands—

$$\frac{\text{Driven gears}}{\text{Driving gears}} = \frac{84}{40}$$

The ratio $\frac{84}{40}$ is broken up:—

$$\frac{84}{40} = \frac{7}{4} \times \frac{12}{10}$$

Then—

$$\frac{7}{4} \times 8 = \frac{56}{32}, \text{ and } \frac{12}{10} \times 4 = \frac{48}{40}$$

Then $\frac{56}{32} \times \frac{48}{40}$ are the gears required, 56 and 48 being the driven, and 32 and 40 the drivers.

The foregoing can be put into words, thus:—

Multiply the lead of the spiral required by 4, because that is the pitch of the lead screw. The number thus obtained will be a numerator, and 40—the number of teeth in the wheel—a denominator of a fraction, which will give the ratio required, that is—

$$\frac{\text{Driven gears}}{\text{Driving gears}}.$$

Break this up into two fractions. Multiply the numerator and denominator by a convenient number or numbers to obtain those corresponding with the number of teeth in the change wheels supplied with the machine. The numbers must be the same for each fraction, but not necessarily the same for both fractions. In fact, it is not often possible to find numbers that will serve for both, and several trials may have to be made to get these. The numbers of teeth corresponding with the numerators will represent the teeth in the driven gears, and the numbers corresponding with the denominators will represent the teeth in the driving gears.

Simple trains of gears can often be used with an intermediate wheel. Thus, to cut a spiral with a lead of 12 inches, the ratio is $12 \times 4 = 48 = \frac{48}{40}$. Then the 48 gear can be put on the worm shaft, and the 40 on the screw, with any intermediate that will make the connection.

When the change gears are obtained, the bed or table has to be set to the angle of spiral required. This is done either by a diagram, or from a table of natural tangents.

If obtained graphically, a diagram of a right-angled triangle is drawn, having one side equal in length to the circumference of the screw to be cut, and the side adjacent equal in length to the lead or pitch, while the hypotenuse, or, strictly, the angle included between the lead and the hypotenuse, is the angle of the spiral. It is by this that the table has to be set.

The angle may be measured, and the spiral bed set to a corresponding angle by the graduations on the clamp bed, or the angle can be marked on the work by a protractor, and the machine set so that the spiral shall be in line with the cutter.

The second method depends on the fact that the natural tangent of the angle of the spiral is the quotient of the circumference of the piece divided by the lead of the spiral. In this method, therefore, the circumference of the piece is divided by the lead. Then note is taken of the number of degrees opposite the figures that correspond with the quotient in a table of natural tangents. Having obtained the angles thus, the spiral bed is set by the graduations on the clamp bed.

Milling Screw Gears.—In considering these gears we must bear in mind the fundamental relations between spiral, helical, and worm gears, which are all alike screw gears. In the first place the only difference between spiral and helical gears lies in the direction of the axes of the wheels engaged. If the axes are parallel the gears are termed helical; if they are at right or other angles, they are spiral. The difference is that in the first case the screw sections must be of opposite hands, or right and left; while in the second they are of the same hand. In the first case also the angles are similar, in the second they are only similar when the axes of the wheels cross at angles of 90°.

With regard to worm gearing, although as a general statement it is true that a worm is a spiral of very short lead, or axial pitch, yet occasional exceptions occur, as when three or four threads are exceeded. The requirements of electric-motor reductions have given a great impetus to the development and expansion of worm driving, so that some many-threaded worms differ in no respect from common spiral gears, excepting in their greater length. Worms of from six to twelve threads are cut successfully and accurately, and worm wheels hobbed to gear with them. Such work lies outside that of the pattern maker, which is seldom practised when worms exceed from two to three threads, and is as a rule confined to single-threaded worms. The growth of the practice of hobbing has enabled good worm gears to be produced, besides which the accurate work of gear-cutting machines permits of the making of gears that have a curious appearance to old-time hands. Examples of this kind occur in the case of worms and their wheels, having say equal numbers of teeth, done to avoid the excessive sliding that would occur in spiral gears. Others are found in the small sizes of worm wheels that are commonly made, having a much smaller number of teeth than the regulation thirty, which is the limit to gears properly proportioned in regard to depth of pitch line, in relation to the length of tooth.

When commencing to design helical gears, the number of teeth or diameters corresponding with velocity ratios are calculated as for spurs, and the circumferences are taken on the faces of the wheels. Then the angle of tooth is considered, and settled (it should not exceed 120°), and cutters selected for the normal pitch (see page 261), which in these wheels scarcely differs from the circular pitch. Double helical wheels are cut as single wheels of opposite hand, and bolted together. As the methods of cutting these do not differ from that of spiral gears, the remarks on the latter (page 261) may be referred to.

The same remark about cutters holds good in respect of worms. The pitch measured along the axis, and the normal, are practically the same in worms of one or two threads, but when these are exceeded a difference becomes apparent.

Cutters are used for the worms only in the best practice, the wheels being hobbed. Exceptions occur in the straight diagonal toothed type of worm wheel, the teeth of which are cut similarly to other spiral or helical gears, since they differ in no respect from these. The cutting of worm gears is treated on pages 274 and 275. We first consider at length the principles and practice involved in cutting spiral gears.

The cutting of spiral gears is one of the duties for which the milling machine is eminently adapted. Gears of large dimensions are not easily tackled, because of the small size of the dividing wheel, and the limitations of the power of the machines. But large gears are not often wanted, and then they can be cast from patterns; or if they have to be cut in quantity, a special type of machine can be built, or specially heavy dividing heads fitted to a stiff machine.

The term spiral gears is used in deference to common usage, though it does not correctly define the form of gear to which it is applied. The term screw gears is correct, but, as a worm gear is as truly a screw gear as the "spiral" form, it is convenient to make the usual distinction in terms. Worm gears and spiral gears differ only in proportions, and not in the principle of their design. A worm helix or thread has a short lead or total pitch, a spiral wheel has this dimension very long by comparison. A worm contains at least one complete revolution of its helix, the spiral wheel usually has but a portion of a revolution of a helix. Worm threads seldom exceed three or four in number, spiral gears may contain portions of ten, twenty, forty, or more threads. The relationship of both to the rack is seen in spiral rack drives, which may be worm, or spiral driven, for slow or rapid speeds. A rack tooth is the basis for both alike in interchangeable gears.

Elements of the Spiral Gear.—Fig. 234 shows the spiral gear in its essentials. P is the primary or axial pitch, p' the normal ditto, and d the circumferential, a the tooth angle of the teeth of a wheel of width of face A.

The case of these gears is not difficult to grasp, if we bear in mind the fact that the pitches of spirals are the bases of velocity ratios. That is, a ratio of 2 to 1 would require a wheel having twice the pitch of spiral of its pinion; and equal ratios pitches of equal length. But only in the single case of equal ratios are the circular pitches equal, because making the ratios and diameters unequal alters the tooth angles, and consequently the circular pitches, though in all cases the normal pitches must be alike. This is where difficulty is often experienced. Moreover, the greater the difference in diameters, the greater will be the differences in the angles, and in the circular pitches, so that the latter will hardly look as though they could be capable of gearing correctly, since one may measure twice or more that of its mating pinion. It follows that the diameters of the pitch lines measured on the wheel faces must have the same ratio as the velocities in spur wheels, to the exclusion of the normal pitch.

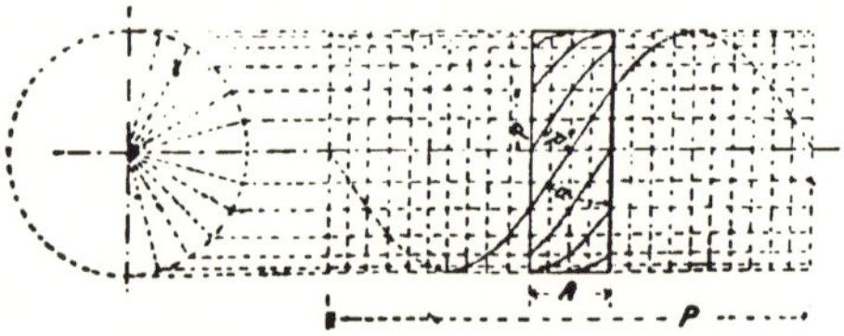

Fig. 234.—Elements of Spiral Gear.

Velocity Ratio.—The first operation necessary in estimating gears is to determine the sizes corresponding with velocity ratio, whether equal or unequal. The rule for this is:—

Divide the centres of the gears by the sum of the terms of the ratio, find the product of twice the quotient by each term separately, and the two products will be the pitch diameters of the two wheels. This rule is not affected by the angles of spirals or worms, but applies to all gears alike.

Example.—Two gears are wanted with a ratio of 2 to 1, the distance between centres to be 6 inches.

Then $2+1=3$

$\frac{6}{3}=2\times 2=4$ in. $\times 2=8$-inch diameter of pitch circle of large wheel.

$2+1=3$

$\frac{6}{3}=2\times 1=2$ in. $\times 2=4$-inch diameter of pitch circle of pinion.

Showing that wheels of 8-inch and 4-inch pitch diameter fulfil the conditions required.

And, having the number of teeth of two gears given, to find the centres :—

Divide half the sum of the teeth of both gears by the pitch.

Say 60 teeth, and 20 teeth of 8 diametral pitch :—

$60 + 20 = 80$

$$\frac{80}{2} = 40 \text{ and } \frac{40}{8} = 5\text{-inch centres.}$$

Spiral gears are used to transmit equal or different velocity

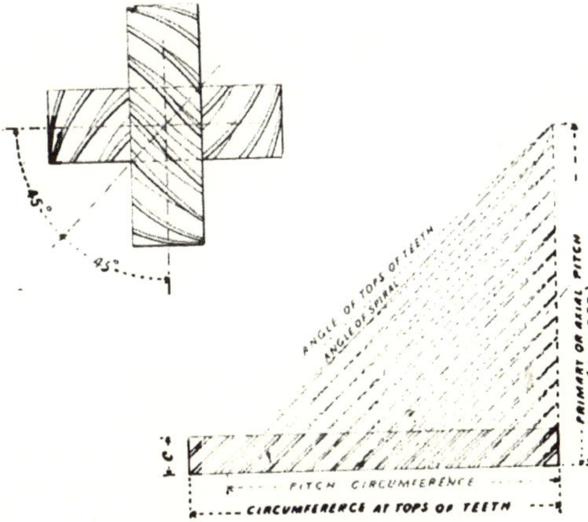

Fig. 235.—Elements of Spiral Gear.

ratios with shafts set at right or other angles, and either between wheels, or between a wheel and rack. In this work puzzles arise in relation to pitch and angle.

In the first place it is well to mention, by way of caution, that the outside diameter or periphery of the blank is not the surface in reference to which calculations refer. These are always taken on the pitch diameter, and the blank (in the diametral pitch system) is always two standard pitches larger than the pitch diameter, just as in spur and bevel gears. The tooth angle at the surface is a steeper one than that at the pitch plane. The cutter, therefore, on its first setting in does not show the true angle of the teeth. The correct angle is obtained by the setting of the table.

Fig. 235 illustrates the relations of the angles of spirals and tops of teeth for the pair of wheels shown.

Pitches.—The pitch of a spiral gear is measured in the plane of its rotation, or parallel with the gear faces. But as the cutter used operates along the plane of the spiral, the result is that cutters have to be selected which do not correspond with the circular pitch, but with the normal pitch, and which is measured at right angles with the plane of the spiral. Also the cutters to be selected must vary with changes in the angles of spirals. But the same cutter must always be used for gears which have to engage together.

The utility of the circumferential pitch is, therefore, as a basis for the measurement of the diameter of the wheel blank, the diameter of which must correspond with the number of teeth, or velocity ratios required, although the cutters must always be selected in reference to the normal pitch.

The angle to which the table carrying the blank has to be set must correspond with the angle of the total pitch or lead of the spiral, of which the wheel teeth are short portions only. The problem, therefore, is to impart one turn to the imaginary cylinder of which the blank forms a short length, while the table is fed along a distance equal to the total pitch or lead.

Angle of spiral is governed by diameter as well as by lead. The smaller the diameter the less the angle, and the larger the lead the less is the angle. It follows that in every pair of gears, excepting those which have their axis at an angle of 45° and their spirals equal, there is no relation common to both wheels except the normal pitch. In wheels of 45° with spiral angles equal, normal, circumferential, axial or primary pitches, and number of teeth coincide, but in all others the first-named only.

Velocity ratios in spiral gears differ from similar ratios in spur gears in an important way. In these last, ratio depends on diameter and number of teeth strictly. That is, a certain number of teeth of a given pitch are inseparable from a certain circumference. But with spiral gears the rule does not follow, except in the case of gears having spirals of 45° of angle. In all other cases the velocity ratios depend on the number of teeth, and not on diameter. Taking an extreme case, that of worm gears, the diameter of a worm exercises no influence on the velocity of the wheel which it drives, but only its number of threads, as one, two, or three. A spiral gear is only a worm of coarse pitch, as also the worm may be considered as a spiral gear of fine pitch.

It follows that by varying the angles of the spirals, in other words, the primary pitches or leads of the helix, it is practicable to make spiral gears in which the ratios will be very different, and greater or less than those which the diameters have (page 271). It is also possible to have coarsely pitched strong teeth on gears of small diameter. In such cases the gear whose circular pitch or angle of spiral is the greater will be the driver, because of the greater obliquity of its teeth.

Graphic Methods.—The problems of spiral gears thus range themselves under two heads, namely, axes at right angles and axes not at right angles, numbers of teeth equal and numbers of teeth unequal.

There are two methods of working out spiral gears—the graphic and the method of calculation. Both are safe, but the last can be obtained more quickly by the assistance of tables. Many, however, feel safer with the graphic.

The following is a concise statement of the graphic method for obtaining angles and teeth of spirals:—

To find the angle of the spiral for a given pitch (axial or primary pitch) and diameter of work, or to find the pitch corresponding to a given angle and diameter, set out a triangle with the pitch as a horizontal line and the perpendicular equal to the circumference of the work, and draw the hypotenuse. The angle of the latter with the horizontal, or pitch length is the angle of the spiral. It is the angle to which the slide of the milling machine is set.

Again, if the angle of spiral is given, draw the perpendicular equal to the circumference of the work, and then draw the hypotenuse at the given angle, to an indefinite length. Complete the triangle by drawing the horizontal until it cuts the slant line. The length of the horizontal so obtained is the primary pitch.

Again, the circumference divided by the pitch gives the tangent to the angle.

To find the pitch, subtract the given angle from 90°. The tangent of the remaining angle multiplied by the circumference gives the pitch.

It must be remembered that in all that has been hitherto stated the pitch surface of the wheels is meant, and not the outside of the wheels. So that to attempt to check the work on the outside by these rules would result in apparent discrepancy. The angle which the cutter would skim over on a trial cut is not that of the pitch surface.

Two equal spiral wheels with right-hand spiral teeth are shown

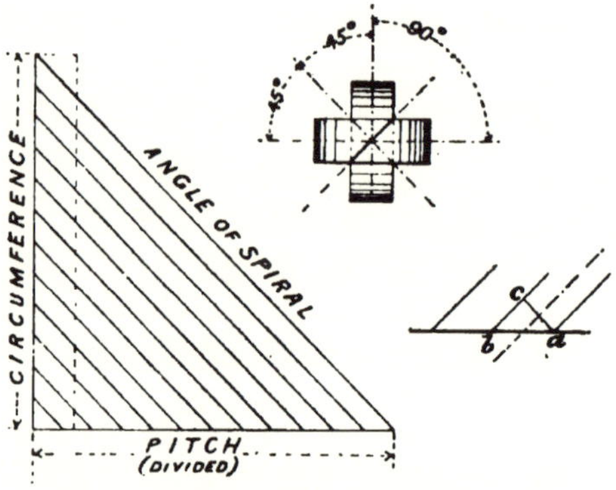

Fig. 236.—Equal Spirals with 45° of Angle.

in Fig. 236, of the kind which most commonly occurs. The angles between their axes are 90°, and the teeth make angles of 45° with these axes and with the wheel faces. The tooth angle is also indicated by a deep line on the upper gear. In order that this angle of 45° shall exist, the pitch or lead of the spiral must obviously be equal in length to the circumference at the pitch line, because any differences in these lengths would not give an hypotenuse of 45°. Stated in another way, the circumference and the pitch of the spiral must be alike, because the tangent of 45° is 1. The development of the wheel is shown to the left, the perpendicular equalling in length the circumference of the wheel, and the length of the base the pitch of the spiral. We lay out the length of the base equal to the circumference, without reference, in this case, to any direct attempt at measurement, because we know such must be correct if the angle of the spiral is to be 45°. Fig. 236 also shows the pitching-out for a 12-toothed wheel. We take no notice yet of the top and bottom of the teeth, because the diameter of the blank gives the first, and the cutter takes charge of the second. The relation between the circumferential pitch *a-b* and the normal *a-c*, in this case (see the diagram to the right, where *a-b* is the hypotenuse of *a c b*), is that of a 45° of angle, being that of the wheel teeth. The circumferential pitch *a-b* can be measured directly; but if calculated, it is obtained for an angle of 45° by multiplying the normal pitch *a-c* by 1·4142, because 1·4142 is the secant of the angle of 45°, or the normal pitch can be divided by the sine of 45°, ·70710, to obtain the same result. Conversely, the circumferential pitch can be multiplied by the sine ·70710 of 45° to obtain the normal.

This difference between the two pitches is very important in determining the selection of cutters for wheels. It is clear that although the circumferential pitch *a-b*, Fig. 236, must be taken for sizing the blanks, the cutters suitable for that pitch, having a thickness of half *a-b* at the pitch line, would cut spaces too wide and leave teeth too thin when cutting normally to *a-c*. Thinner cutters must therefore be selected than those used for spur wheels of the same nominal pitch. Hence the rule:—As the angle *a b c* equals the angle of the spiral, and the line *b-c* corresponds with its cosine, multiply the cosine of the angle of the spiral by the circumferential pitch. The product will be the normal pitch, one half of which will be the thickness of the cutter at the pitch line.

The reverse of this rule is to be borne in mind also:—

Divide the normal pitch by the cosine of the angle of the spiral. The product will be the circumferential pitch. (*N.B.*—"Circumferential" and "circular" pitch are terms used here loosely to denote the same pitch—that in the plane of the wheel faces as distinguished from the normal, and not necessarily in contrast to diametral pitch.) The diametral pitch of the gear can be obtained, after the selection of the cutter, by multiplying the diametral pitch of the latter by the secant of the angle of the tooth. A useful table is given overleaf for different pitches and angles.

TABLE OF SPIRAL GEAR—DIAMETRAL PITCHES.

Diametral Pitch of Cutter.	Depth to be cut in Gear.	Angle—15°. Corresponding Diametral Pitch of Spiral Gear.	Angle—30°. Corresponding Diametral Pitch of Spiral Gear.	Angle—45°. Corresponding Diametral Pitch of Spiral Gear.
	Inch.	Inch.	Inch.	Inch.
24	·090	·0431	·0481	·0589
22	·098	·0470	·0525	·0643
20	·108	·0517	·0577	·0707
18	·120	·0575	·0641	·0785
16	·135	·0647	·0721	·0883
14	·154	·0739	·0824	·1010
12	·180	·0862	·0962	·1178
11	·196	·0941	·1049	·1285
10	·216	·1035	·1154	·1414
9	·240	·1150	·1283	·1571
8	·270	·1294	·1443	·1767
7	·308	·1479	·1649	·2020
6	·359	·1725	·1924	·2357
5	·431	·2070	·2309	·2828
4	·539	·2588	·2886	·3535
3	·719	·3451	·3849	·4714
2	1·078	·5176	·5773	·7071

The diameter of the blank, equalling two pitches, added to the pitch diameter must be calculated on the basis of the normal pitch for which the cutter is selected, and not on the circumferential pitch, and the same remark applies to the depth of tooth. And hence, too, the circumference of a gear blank equals the number of teeth multiplied by the normal pitch, multiplied by the secant of the angle of spiral, plus two standard pitches.

Or,

$$\text{Circ.} = \text{number of teeth} \times \text{normal pitch} \times \text{secant } a + 2p.$$

Also, the pitch or lead of the spiral and the circumference of the wheel stand in the relation of secant and cosecant of the same angle of spiral. Hence the lead of the spiral equals the number of teeth multiplied by the normal pitch, multiplied by the cosecant of the angle of the spiral.

Or,

Pitch, or lead of spiral = number of teeth × normal pitch × cosecant a.

If a piece of paper is cut, as in Fig. 237, and wrapped round a cylinder of the same diameter as the pitch plane of the spiral gear in Fig. 236, but considerably longer than the thickness of the wheel, we shall have the result shown in Fig. 237, in which the primary pitch or lead of the spiral is also the base of the triangle, the circumference is the same as the perpendicular, and the angle of spiral is the hypotenuse.

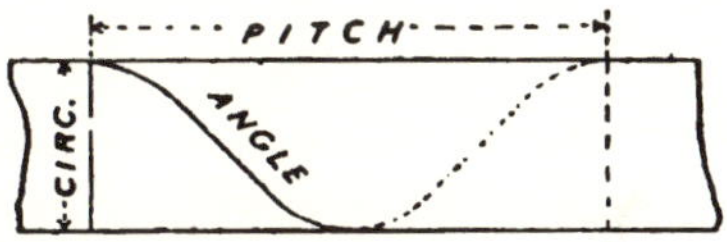

Fig. 237.—Triangle wound round a Cylinder.

Also in Fig. 236 the spiral wheel is shown developed along the

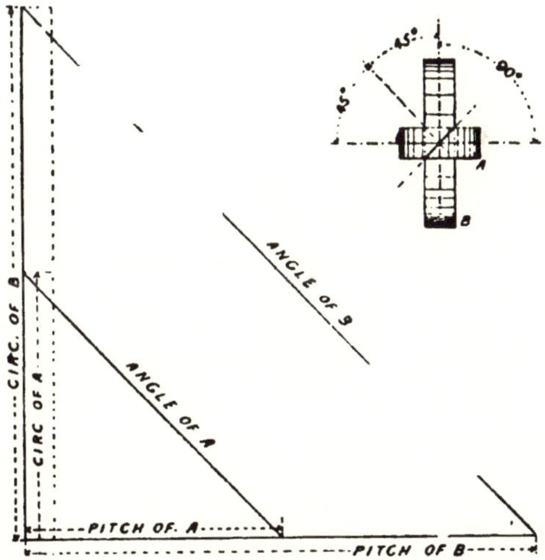

Fig. 238.—Spiral Gears of Different Diameters, but Equal Angle.

edge of the circumference, from which the fact is clear that a spiral rack is only a developed spiral wheel, and hence such gears will engage correctly.

Going a stage further, and keeping still to wheels the axes of which are at right angles, the question of unequal velocity ratio arises. The velocity ratio can be altered in gears at right angles by manipulating the numbers of teeth and retaining the same angles of spiral. The pitch diameters vary exactly as the velocity ratios, and either wheel or pinion can drive or be driven just as in the previous example. The relations of gears with 2 to 1 ratio are shown in Fig. 238, and a comparison of the diagrams illustrates the fact that the length of the spirals and the diameters of the cylinders are exactly 2 to 1 to produce equal angles of 45°. The angle of the tangent line being equal, the normal and circumferential pitches would be alike in each wheel.

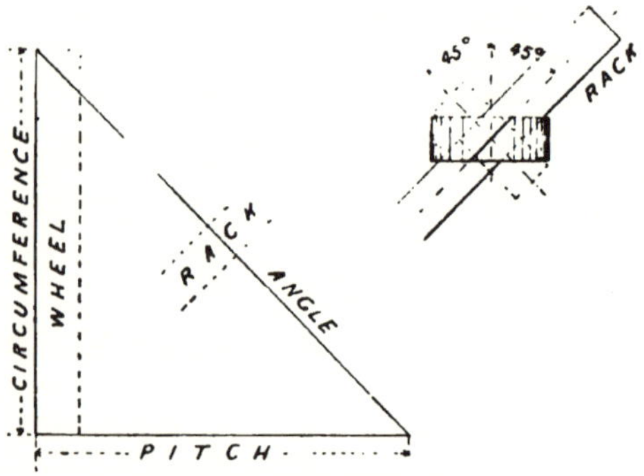

Fig. 239.—Spiral Gear, and Rack.

Again, one spiral wheel may be imagined to be spread out flatwise, and then we have a spiral rack, Fig. 239, which the spiral wheel will drive. The normal and circumferential pitches here are alike in the rack, provided, as before, the common tangent stands at an angle of 45° with the axes. The wheel outlined in Fig. 239 would be impracticable with a 45° tooth angle, but such a gear would be possible by altering the tooth angles.

It is practicable to impart unequal velocity ratios to spiral gears of the same diameters and with their axes at right angles; but then neither of the spirals will be of 45° angle, but each will have an angle different from the other. Then the wheel with the larger number of teeth will be based on a spiral having the same relation to the other as the ratio required. If the ratio is as 2 to 1, then the ratios of the length of spirals will be as 2 to 1. The line of action or tangent line common to the two spirals and passing through the pitch point will divide the axes in the relation of 2 to 1, Fig. 240. The angles of the two spirals are shown projected in Fig. 240, with the outlines of the developed wheels indicated.

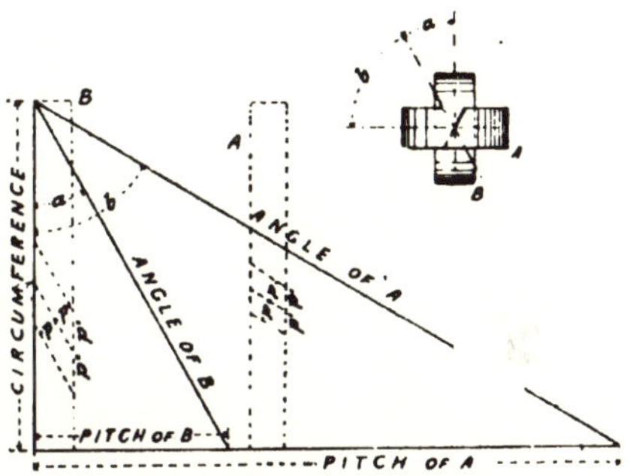

Fig. 240.—Spiral Gears with Teeth of Different Angles.

In gears with unequal numbers of teeth, that having the most acute angle, or the spiral of shorter primary pitch, or the coarser circular pitch, is the driver.

Looking at the diagram, Fig. 240, of the developed wheels, we see that the circular pitches p', p' differ by a considerable amount, though the normal pitches p, p are alike. The pitches can be obtained in this way, though a simple calculation gives them readily, see page 261.

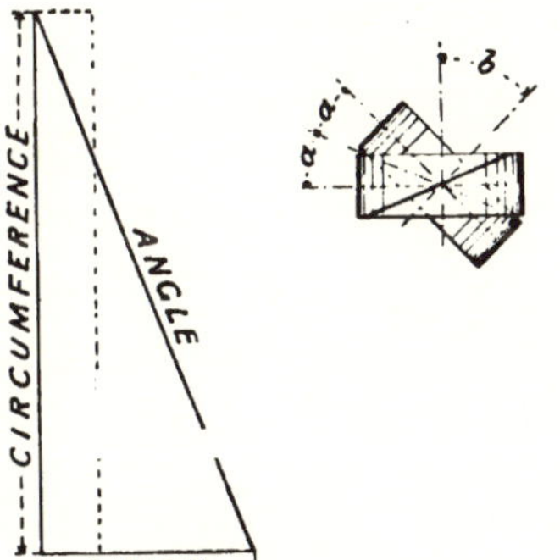

Fig. 241.—Spiral Gears with Teeth of Same Angle.

Spiral gears with their axes not at right angles may be geared in equal or unequal ratios. If equally, the angle included between the axes, or, what is the same thing, the angle between the middle plane of the wheels, may be bisected equally, and the line of bisection will coincide with the tangent line common to the teeth of both wheels, and passing through the pitch point, Fig. 241, where a, a are equal and their sum equals b. The development of Fig. 241 is shown to the left, and here the circular pitches are alike in both wheels.

Wheels with their axes set at an angle must be bisected unequally for unequal ratios, Fig. 242, in which case the spirals are of different angles and the circular pitches different. The development is shown in Fig. 242 for wheels A and B.

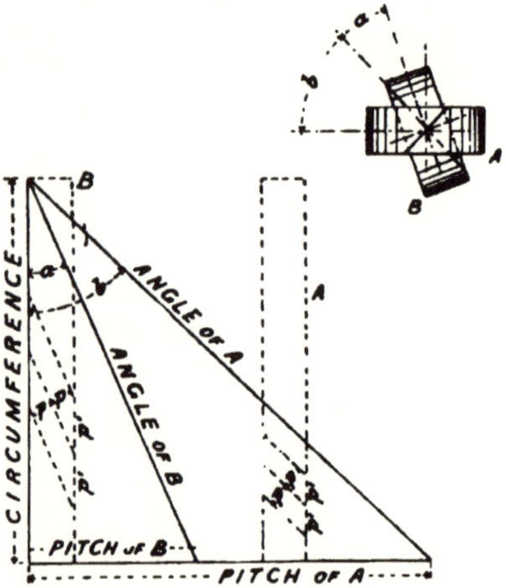

Fig. 242.—Spirals with Unequal Ratios.

Racks or spur gears can be mated with spiral wheels at an angle. In such cases the spirals must be the drivers. In Fig. 239 a spiral wheel with its teeth set at an angle of 45° was shown engaging a rack, or a spur wheel might be substituted for the rack. In Fig. 243 a spiral with an angle of 45° engages a spur wheel. In both these cases the pitch of the rack or spur is the same as the normal pitch of the spiral gears. This helps to illustrate the reason why cutters must be selected based on normal pitch for any spiral gears irrespective of the differences in circular pitch. In these examples, the shading only indicates the circular forms of the gears.

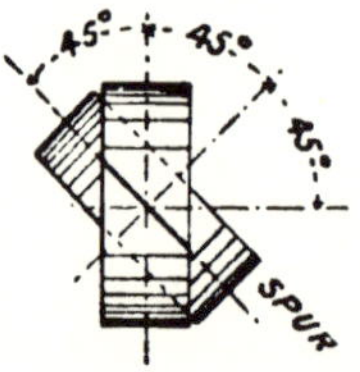

Fig. 243.—Spiral Gearing with Spur.

Trigonometrical Rules.—Many prefer to employ these graphic methods, which are safe. But they may be dispensed with by using a few of the simple rules of plane trigonometry. Since the functions of angles are constant, the relations of circumference, pitches, angles of spirals, and normal and circular pitches are obtainable. These will be first briefly explained, and then formula for spiral gears based thereon tabulated.

Fig. 244 illustrates the functions of a right-angled triangle, which are applicable to spiral and bevel gears having axes at right angles.

A B C represents the triangle, in which A C is the base, B C the perpendicular, and A B the hypotenuse. In this we recognise also the base as equivalent to the pitch of a spiral, the perpendicular as equal to the circumference of the cylinder, and the angle of the hypotenuse as the same as the angle of the spiral. If a piece of paper is cut to the shape of the figure and wound round a cylinder of the same circumference as the length of the perpendicular, the spiral will be developed on the cylinder.

An arc is shown struck from C to D from A as centre. The perpendicular C B is the tangent to this arc, and it is perpendicular to the base, or radius A C which meets it at the point of tangency. The tangent C B is also the tangent to the angle C A B, and is used in calculation. If the length of the perpendicular or tangent C B is known, the number of degrees in the angle C A B can be found in a table of tangents, and this, as we see, is equivalent to the angle of spiral.

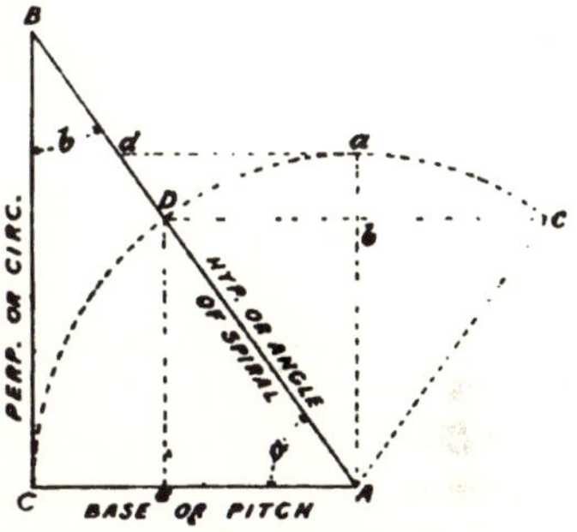

Fig. 244.—Trigonometrical Functions.

There are two angles in any right-angled triangle, either of which may be required to be known. Hence, when speaking of the angle C A B, the base C A is termed the side adjacent to that angle, and, when speaking of the angle C B A, the perpendicular C B is the side adjacent to that angle.

The following rules are based on the functions of right-angled triangles:—

To find the tangent of either acute angle in a right-angled triangle—

Divide the side opposite the angle by the side adjacent the angle, and the quotient will be the tangent of the angle. In Fig. 244 the rule might read correctly:—Divide the perpendicular by the base, and the quotient will be the tangent of the angle C A B.

Angles are measured by arcs in degrees and minutes. Thus *a* is the measure of the angle C A B, and *b* that of the angle C B A.

The complement of an arc is the difference between it and 90°. Thus *a* D in Fig. 244 is the complement of C D, and *vice versa*.

The supplement of an arc is the difference between it and 180°, or of a semi-circle. The sine of an arc is the line drawn from one extremity of the arc perpendicular to the diameter passing through the other extremity. Thus *b* D is the sine of the arc *a* D, and it is equal to half the chord of the arc D *c*. The versed sine is the length *a b* between the chord and the arc. A *d* is the secant of the arc *a* D.

The complements of the functions of an arc are denoted by the prefix co—as cosine, cotangent, and cosecant. Thus the arcs C D, D *a* being the complement of each other, the sine, tangent, and secant of either is the cosine, cotangent, and cosecant of the other. Thus *b* D, the sine of *a* D, is the cosine of D C. D *e*, the sine of D C, is the cosine of *a* D; *a d*, the tangent of *a* D, is the cotangent of D C; C B, the tangent of C D, is the cotangent of *a* D; A *d*, the secant of *a* D, is the cosecant of D C; and A B, the secant of D C, is the cosecant of *a* D. In any arc *a* D, therefore, the versed sine *a b* and cosine D *e* make up the radius A *a* or A D. And the radius A *a*, the tangent *a d*, and the secant A *d* make the right-angled triangle A *a d*. Also the radius A C, the cotangent C B, and cosecant A B make the other right-angled triangle A C B.

The sine of an angle is the sine of the arc that measures the angle. It is always inside the arc, and can never be longer than the radius. As the arc approaches 90°, the sine comes nearer to the radius, or 1. Thus (*a*), the sine and cosine can never be greater than unity; and (*b*), the secant and cosecant can never be less than unity; while (*c*), the tangent and cotangent can have any value between zero and infinity.

Referring again to Fig. 244, the trigonometrical ratios of the angle at A may be summarised as follows:—

$$\text{The sine of A} = \frac{\text{perpendicular}}{\text{hypotenuse}} = \frac{\text{B C}}{\text{A B}}$$

$$\text{The cosine of A} = \frac{\text{base}}{\text{hypotenuse}} = \frac{\text{A C}}{\text{A B}}$$

$$\text{The tangent of A} = \frac{\text{perpendicular}}{\text{base}} = \frac{\text{B C}}{\text{A C}}$$

$$\text{The cotangent of A} = \frac{\text{base}}{\text{perpendicular}} = \frac{\text{A C}}{\text{B C}}$$

$$\text{The secant of A} = \frac{\text{hypotenuse}}{\text{base}} = \frac{\text{A B}}{\text{A C}}$$

$$\text{The cosecant of A} = \frac{\text{hypotenuse}}{\text{perpendicular}} = \frac{\text{A B}}{\text{B C}}$$

We have now the following trigonometrical rules for spirals. In Fig. 244 the circumference of the cylinder around which the spiral is wound corresponds with the perpendicular of the triangle, or the side opposite the angle of the spiral.

In words the rules are:—

To obtain the angle of the spiral—

Divide the perpendicular (circumference of cylinder or spiral) by the base (number of inches of spiral to one turn), and the quotient will be the tangent of angle of spiral.

When the angle of spiral and circumference are given, to find the pitch—

Divide the perpendicular (circumference of cylinder or spiral) by the tangent of angle, and the quotient will be the base (pitch of the spiral).

When the angle of spiral and the pitch of the spiral are given, to find the circumference—

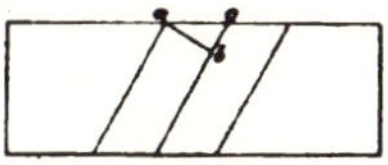

Fig. 245.—Circular and Normal Pitches.

Multiply the tangent of angle by the pitch, and the product will be the circumference (perpendicular).

The angles which the circular and normal pitches make with each other bear the same relation as the axes of the wheels and the tooth angles. Hence they can be deduced one from the other trigonometrically as well as by direct measurement. In Fig. 245 *a c* is the circular pitch, and *a b* the normal, and the angle *b a c* is the same as that of the spiral, *a b* is the cosine of the angle *b a c*.

Then—

$$\text{Circular pitch} \times \text{cosine angle} = \text{normal pitch}.$$

and—

$$\frac{\text{normal pitch}}{\text{cosine angle}} = \text{circular pitch}.$$

We have seen (page 264) that the ratios of diameters are alike in some instances, and different in others. In the latter case the condition to be observed is that the pitch of the spiral of each gear shall be equal to the circumference of the other, or any fraction of the same, depending on the ratio desired. The velocity ratio is measured by the number of teeth, and not by the diameters. The object first is to select the angles, and from these the corresponding diametrical pitches.

Example.—Required two gears at right angles, measuring respectively 2-inch and 4-inch diameter, the 4-inch gear not to be used for speeding up the 2-inch, but for speeding it down, in the ratio of 2 to 1. The circumference of 4 inches is 12·5664, that of 2 inches is 6·28319. Half the circumference of the 2-inch gear is 3·14159, which is the lead of the spiral of the 4-inch gear. But the lead of the spiral of the 2-inch would be twice the circumference of the 4-inch gear (the axes being at 45°) or 25·1328 inches. Dividing the circumference by the pitch gives the tangent of the angle. Hence—

Circumference of 4-inch = 12·5664.

And $\frac{12{\cdot}5664}{3{\cdot}14159} = 4$, and 4 = a tangent of angle of nearly 76°. 76° from 90° leaves 14°, which must be the angle of the 2-inch gear. These relations are shown in the diagram, Fig. 246.

To obtain the cutters by the rule (page 261):—

The secant of 76° is 4·1336. That of 14° is 1·0306. Selecting an 8-pitch cutter, and multiplying this by the secant of the angle, we have:—

4·1336 × ·1250 = ·5167 = the corresponding diametrical pitch for 76°. And 1·0306 × ·250 = ·1288, that for 14°.

The numbers of teeth that will work out nearest to the sizes of wheels required are:—

24 for the 4-inch wheel—

24 × ·5167 = 12·4008 circ.

And, 48 for the 2-inch wheel—

48 × ·1288 = 6·1824 circ.

So that diameters would have to be slightly modified, and the leads of the spirals corrected to suit.

Another way of putting it is this:—

$$\text{The tangent of the spiral of wheel} = \frac{\text{pitch diam. of wheel} \times \text{number of teeth in pinion}}{\text{pitch diam. of pinion} \times \text{number of teeth in wheel}}$$

and

$$\text{The tangent of the spiral of pinion} = \frac{\text{pitch diam. of pinion} \times \text{number of teeth in wheel}}{\text{pitch diam. of wheel} \times \text{number of teeth in pinion}}$$

The illustration given, Fig. 246, is an extreme case, and it is not a desirable gear, but it shows the flexibility of the spiral gears.

When axes are at any angle. When the numbers of teeth in the two wheels are given the angles are found by a graphic method. In Fig. 247, *a b* and *c b* are the axes of gears which include the angle *o* between them. On these axes distances are set off *b a*, *b c* representing the velocity ratios, and a parallelogram completed, so that the number of the teeth in the wheel : the number of teeth in the pinion :: *a d* : *c d*. The tangent *b d*, if common to the teeth passing through the point of contact, ensures the least amount of sliding. But if the angle *o* is bisected at *e b*, the end thrust is equally distributed on both shafts. If an angle is taken, *b f* bisecting *e b d*, a slight advantage of each kind is gained.

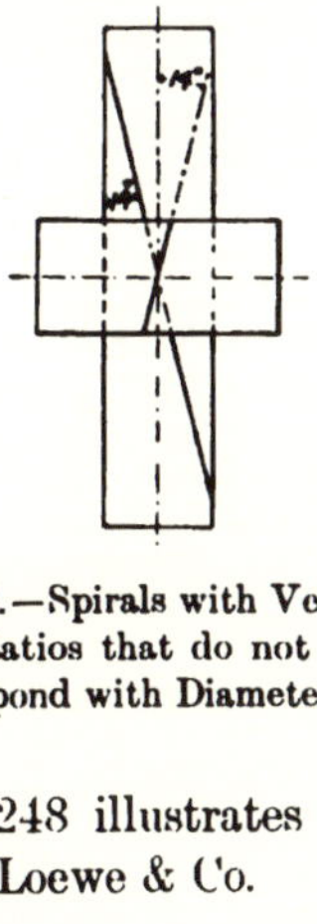

Fig. 246.—Spirals with Velocity Ratios that do not correspond with Diameters.

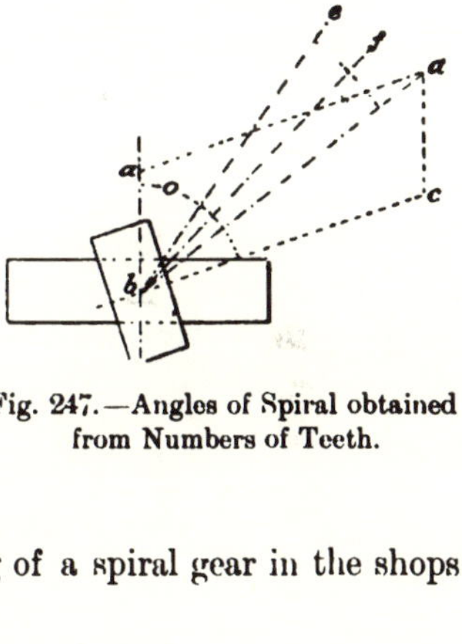

Fig. 247.—Angles of Spiral obtained from Numbers of Teeth.

Fig. 248 illustrates the milling of a spiral gear in the shops of Ludwig Loewe & Co.

Worm Gears.—There is more trouble experienced in getting good worm gears than spurs, or even bevels. I have seen some so badly cut that the wheel teeth have had to be trimmed and filed to make a decent fit, and have also known gears, machine-moulded from pattern blocks, substituted for those which have been cut, and better results obtained thereby. There are three reasons for this poor practice: one is that special types of machines are best adapted for cutting the concave forms of worm-wheel teeth; that a rather costly hob is generally necessary; and, in the case of small wheels, the conditions required to ensure good gear are not always understood.

A worm wheel in which the teeth are straight, and inclined at an angle corresponding with the angle of the thread of the worm, is easily cut with a rotary cutter, and it is sufficiently good for a large quantity of work. Wheels used for dividing purposes, and for light running, are mostly made like this. But when heavy work has to be done, and good wearing capacity is desired, it is

Fig. 248.—Milling Spiral Gear.

the usual practice to make the teeth envelopes of the worm. Such gears, when run at moderate speeds, are very durable. Good results are obtained when both wheels are in cast iron, provided they are well lubricated with plumbago and grease. Sometimes steel worms run with cast-iron wheels, because iron worms always wear out before their wheels. The best results are found when the worm is of phosphor bronze running in an iron or steel wheel.

The breadth of the teeth of a worm wheel is about half the diameter of the worm, though it is often less. There is no advantage in excessive width. Three complete turns of a thread are sufficient for contact in any case. The diameter of a worm is generally from four to five times the circular pitch. Worm-wheel teeth are reckoned by circular instead of diametral pitch. The pitch is measured along the axis of the worm and in the middle plane of the wheel, and is therefore axial. Normal pitch, though more correct, would not be so convenient a working basis. There is a way of reckoning pitch which differs from that of common gears, and which gives it as so many threads in the linear inch, as 1 pitch or 2 pitch.

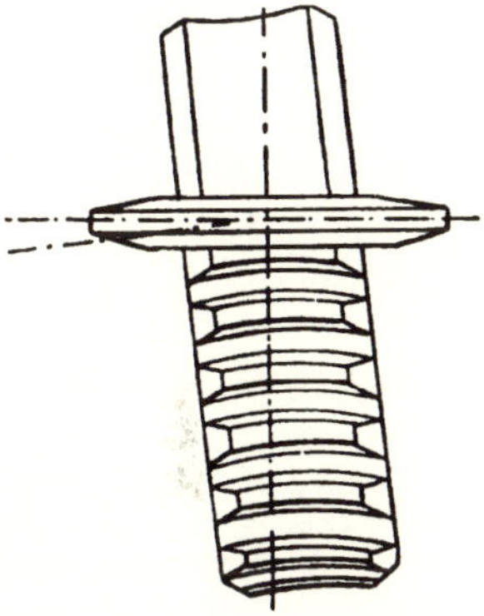

Fig. 249.—Cutting Worm Wheel Teeth with Common Cutter.

Worm wheels frequently have their concave teeth cut with rotary cutters, Fig. 249, on milling machines. The results are satisfactory as far as pitching and tooth angle are concerned, but the curved teeth are not true envelopes of the worm, as they are when wheels are hobbed. There is doubtless more surface contact between the teeth than when straight teeth are employed, but the advantage is probably not very great.

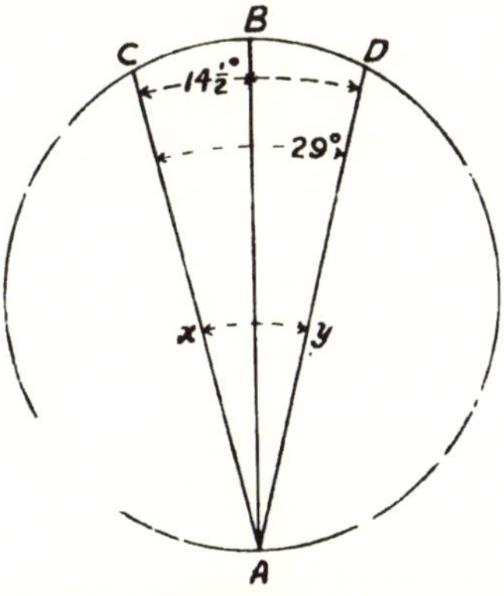

Fig. 250.—Obtaining the Shape of Tool for Cutting Worms.

When teeth are cut thus, the wheel blank is carried between the index centres, and the centre of the wheel brought under the centre of the spindle which carries the cutter. The table is set with the saddle brought to the angle of the teeth, and the vertical feed of the table takes the cutter into the proper depth.

Cutting wheel teeth to become true envelopes of the worm

can only be done properly by a hob in shape like the worm, or by a machine in which a single cutter is presented at various angles. This requires a stiff machine, support to the wheel rim against the cutting, and a motion of the wheel blank relatively to the hob at the same rate as the velocity ratios of the wheels when finished.

The involute tooth is accepted as the proper form of worm gear, not because cycloidal forms cannot be cut, but because it simplifies matters to turn or mill the worm thread with a vee-shaped tool or cutter of definite angles, and to make that—the rack tooth—the basis for all wheels.

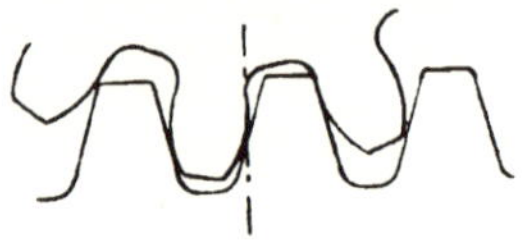

Fig. 251.
Worm Gear Teeth.

Though the threads of worms are very commonly turned in the lathe, yet they are also done more properly in the milling machine, using a swivel head set to the proper angle, and a rotary rack cutter.

Turning a worm thread is just like turning any screw thread in the lathe, whether it be single, double, or treble threaded. Fig. 250 shows how the shape of the tool for cutting the rack-shaped teeth is obtained. Upon a line A B draw a circle A C, B D. From B lay off the distance B C and B D, each equal to one quarter the diameter of the circle. Draw C A, D A, which make an angle of 29°—the angle of inclination of the sides of the rack tool. Make the breadth of the tool at the end $x-y$ equal to 0·31 of the circular pitch. The width of the top of the thread of a worm is 0·335 of the circular pitch. The relations of the worm and wheel teeth are shown in Fig. 251.

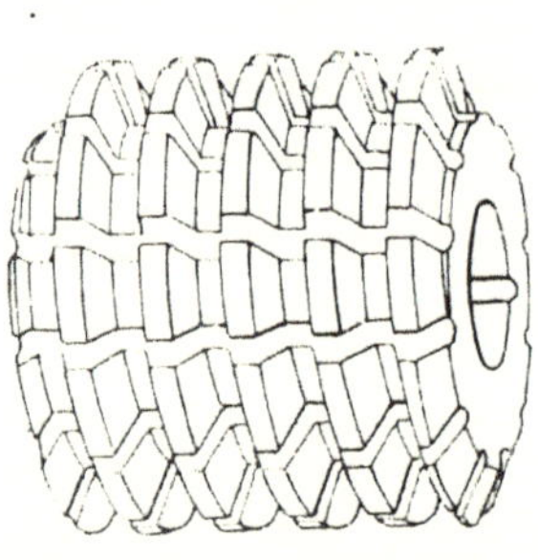

Fig. 252.—Hob for Worm Wheel.

The hob, Fig. 252, by means of which the wheel teeth are cut, is turned in steel to the same shape as the actual worm, plus a little extra in diameter, sufficient to give the bottom clearance between worm and wheel. The proportions of the hob should be these:—The diameter exceeds that of the worm by twice the amount of bottom clearance in the wheel teeth. The depth of the hob thread is equal to that of the working depth, plus the clearance; the diameter at the bottom of the hob thread is the same as that of the bottom of the worm threads. Grooves are then slotted out in order to form teeth, which are backed off and hardened, converting it into a cutting tool. The grooves are planed or milled out with a round-edged cutter, the straight portion terminating at the bottom of the thread, leaving the concavity below the bottom.

When a wheel has its teeth cut in this way, every portion of each tooth comes into successive contact with the thread of the worm during the revolution of the latter. At no instant is a tooth face a perfect envelope of the worm thread, and therefore any attempt to cut such teeth perfectly with a single rotary cutter, as in Fig. 249, must needs result in a tooth space of uniform section, which would not give a correct gear. Neither teeth sections nor spaces are uniform and symmetrical, and the departure from symmetry is more pronounced with multiple-threaded than with single-threaded worms. The forms obtained result from the peculiar relations of the worm and the concave section of the wheel rim, and the effect is exactly as though the worm cut out its wheel teeth in a softer substance. The wheel blank is rotated through change wheels, regulated to move it at the proper velocity ratio in relation to the worm. This is essential to ensure perfect results, notwithstanding that a worm will cut its wheel teeth without change gears. To have the pitch right, however, by this method it is first necessary to block out the teeth; because if this precaution is not taken—that is, if the hob cuts its way into the blank from the beginning,—the first cutting will take place on a diameter larger than the pitch diameter, and by the time the wheel is finished the pitch on the pitch diameter will be less than that of the worm, and the teeth too many in number.

A relieved hob is formed on the same principle as the relieved milling cutters, so that its sectional form does not alter by regrinding on the front faces of the teeth.

J. E. Reinecker, one of the leading toolmakers in Germany, cuts worm wheels with a hob shaped as in Fig. 253, to which an endlong movement is imparted, as shown in its successive advances in Fig. 254, during its revolution. The advantage is that the hob has not to be fed deeper into the blank as it cuts, but it is set to correct centres once for all. The blank is mounted on an arbor in a machine illustrated on pages 147 and 148.

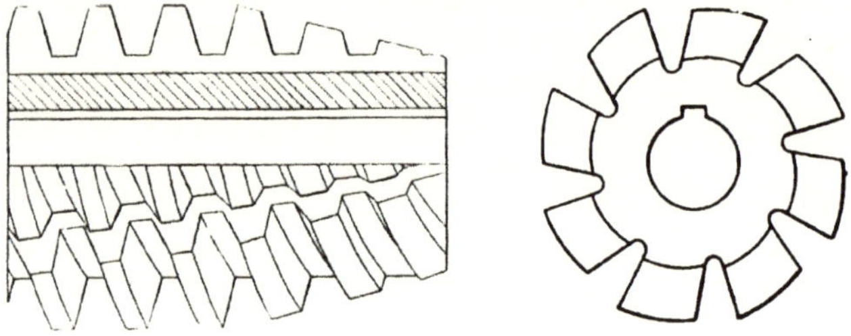

Fig. 253.—Triple-threaded Tapered Hob.

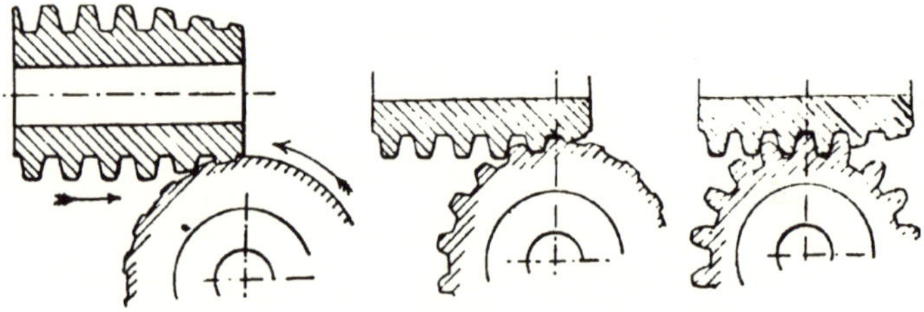

Fig. 254.—Reinecker Hob Cutting Worm Wheel.

TABLE OF WORM GEAR—DIAMETRAL PITCHES.
Brown & Sharpe Standard Depth.

Pitch of Worm.	Corresponding Diametral Pitch of Gear.	Working Depth of Worm Thread.	Pitch of Worm.	Corresponding Diametral Pitch of Gear.	Working Depth of Worm Thread.
Inches.	Inch.	Inch.	Inches.	Inch.	Inches.
1- 8	·0398	·0796	$\frac{5}{8}$	·1988	·3978
1- 6	·0531	·1062	$\frac{3}{4}$	·2388	·4774
1- 5	·0637	·1274	$\frac{7}{8}$	·2785	·5570
1- 4	·0796	·1592	1	·3183	·6366
1- 3	·1061	·2122	$1\frac{1}{4}$	·3982	·7960
3- 8	·1193	·2388	$1\frac{1}{2}$	·4774	·9550
4-10	·1273	·2546	$1\frac{3}{4}$	·5573	1·1141
1- 2	·1592	·3184	2	·6366	1·2732

The depths given do not include the clearance.

Rule.—Multiply the corresponding diametral pitch by the number of teeth to get the pitch diameter, and add the working depth of the tooth to this to get outside diameter.

Example.—40 teeth, 4-pitch worm. ·0796 inch × 40 = 3·184 inches pitch diameter. To the pitch diameter add the working depth of thread, ·1592 inch, which gives 3·343 inches as outside diameter of flat top gear, or throat diameter of curved face gear.

There are indications that the practice of milling screws instead of cutting them in the lathe will assume great importance in the machine shop. The sizes produced are constantly increasing, and considerable economies effected over lathe work. The two best known machines, the Liebert, and the Pratt & Whitney, are constructed in a number of sizes, and produce both square and vee-threaded screws, the cutters being of special type. In the Liebert, cutters of pressed sheet steel are employed. The machine has a long bed, carrying a head which supports the blank, the latter being fed by change gears situated at the end of the bed. The cutter head has adjustment for height, and a swivel motion, to bring the cutter into correct angling with the screw being operated upon.

In the Pratt & Whitney a somewhat different design is made, the screw being carried between a head and a tailstock, in lathe fashion, while the cutter slide travels along the bed, and has a swivelling head for the cutter arbor. The largest Pratt & Whitney machine takes up to 12 in. diameter by 48 in. in length, being therefore suited for the production of large worms, spiral gears, etc.

CHAPTER X.

SPUR AND BEVEL GEARS.

Spur and Bevel Gears — Diametral Pitch — Blanks — Cutters — Projection of Bevel Wheels — Multiple Cutters — Multiple Centres — Milling Squares — Tapered Work.

Spur and Bevel Gears.—The diameters of spur and bevel wheel blanks, in common with spirals and worm wheels are determined by the amount of addendum added to pitch diameter. This is generally equal to one diameter pitch. Thus in wheels of 4-pitch the addendum would measure ¼ inch, because there are four teeth for each inch of diameter in a 4-pitch wheel. This is a simple rule, and easily borne in mind and applied. It simply means that the blank has to be turned to pitch diameter plus two diameter pitches.

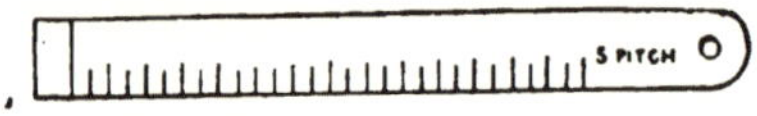

Fig. 255.—Gear Tooth Rule.

To save risk in making even so simple a calculation some firms make specially graduated rules, divided into 18ths, 20ths, 22nds, and so on, Fig. 255. Then, to size a wheel, the number of teeth is taken on the line of corresponding divisions. If a 20-pitch, say, the number of divisions corresponding with the number of teeth is read off on the line of 20ths, and two of the 20th divisions added thereto gives at once the total diameter of the blank.

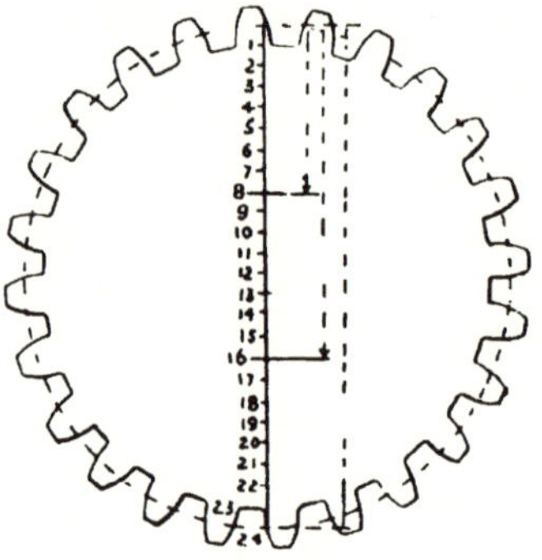
Fig. 256.—Diametral Pitch.

The use of diametral pitch, illustrated by the diagram in Fig. 256—an 8-pitch wheel—simplifies matters in calculation. It is very easy to lay down numbers of simple calculations (which need not be given here) all referable to this, and all facilitating the work of gear cutting, while, when necessary, diametral and their corresponding circular pitches can be ascertained by dividing 3·1416 by the diameter pitch for the circular pitch, and the latter can be converted to diametral by multiplying it by 0·3183, which is a short way of dividing by 3·1416.

TABLE OF DIAMETRAL PITCH, WITH ITS EQUIVALENT CIRCULAR PITCH IN THE ADJOINING COLUMN.

Diametral Pitch.	Circular Pitch.	Diametral Pitch.	Circular Pitch.	Circular Pitch.	Diametral Pitch.	Circular Pitch.	Diametral Pitch.
				Inch.		Inch.	
2	1·57	11	·280	$1\frac{3}{4}$	1·79	$\frac{3}{4}$	4·19
$2\frac{1}{4}$	1·39	12	·262	$1\frac{1}{2}$	2·09	$\frac{11}{16}$	4·57
$2\frac{1}{2}$	1·25	14	·224	$1\frac{7}{16}$	2·18	$\frac{5}{8}$	5·03
$2\frac{3}{4}$	1·14	16	·196	$1\frac{3}{8}$	2·28	$\frac{9}{16}$	5·58
3	1·05	18	·174	$1\frac{5}{16}$	2·39	$\frac{1}{2}$	6·28
$3\frac{1}{2}$	·898	20	·157	$1\frac{1}{4}$	2·51	$\frac{7}{16}$	7·18
4	·785	22	·143	$1\frac{3}{16}$	2·65	$\frac{3}{8}$	8·38
5	·628	24	·130	$1\frac{1}{8}$	2·79	$\frac{5}{16}$	10·06
6	·524	26	·120	$1\frac{1}{16}$	2·96	$\frac{1}{4}$	12·56
7	·448	28	·112	1	3·14	$\frac{3}{16}$	16·75
8	·392	30	·104	$\frac{15}{16}$	3·35	$\frac{1}{8}$	25·12
9	·350	32	·098	$\frac{7}{8}$	3·59	$\frac{1}{16}$	50·24
10	·314	...	...	$\frac{13}{16}$	3·86	...	...

Blanks.—When sizing the blanks of bevel and mitre wheels the safest course is to draw them carefully in section, Fig. 257, adding, as in spurs, one diameter pitch beyond the pitch line on the large diameter. By drawing all lines in section the dimensions at the small end are obtained, and the working depth of tooth and the thickness of tooth or tooth space, which are equal, on the small end. The working depth is of course equal to the addendum. The bottom clearance, added below this, is generally equal to one-tenth the thickness of the teeth on the pitch line, measured on the large side. The thickness of a tooth on pitch line, leaving no flank clearance, is half the pitch. It is obtained in diameter pitches by dividing 1·57 by the diametrical pitch. In a 5-pitch it would be $\frac{1 \cdot 570}{5} = 0 \cdot 314$, and a tenth of this would be

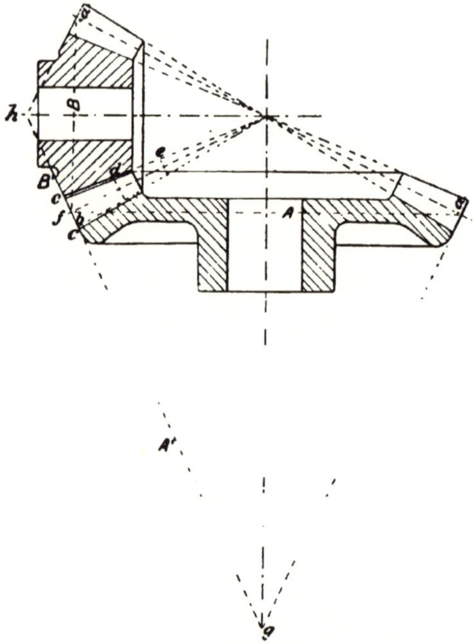

Fig. 257.—Development of Bevel Gears.

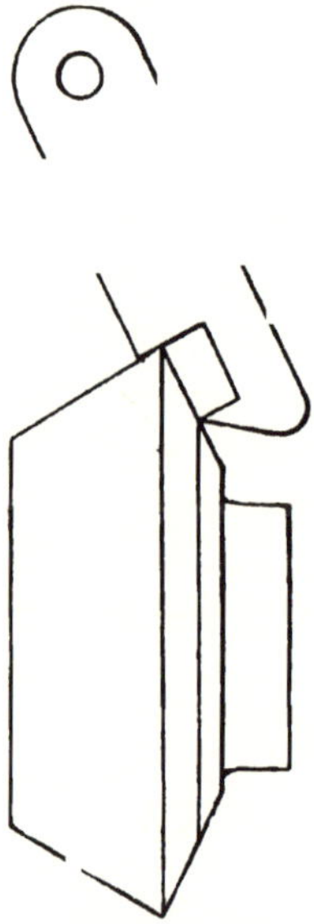

Fig. 258.—Gauge for Tooth Length.

0·031 bottom clearance for a 5-pitch tooth. Tables are used in shops.

The blank, Fig. 258, is turned to the dimensions taken from the drawing, Fig. 257, and it is well to run round two lines taken from the drawing giving the depth of tooth to be cut, Fig. 258, using a gauge for the purpose, one of which is supplied for each pitch by firms who manufacture cutters, &c. The depth of bottom clearance *c*, Fig. 257, is generally made alike at large and small ends in wheels which are cut with rotary cutters.

There is no extra trouble involved in striking-out wheels which work at angles other than right angles, or wheels of different diameters. The methods illustrated in Fig. 257 show how the bounding faces of the teeth converge to the apex of the cones, and that the ends of the teeth stand at right angles with the pitch planes.

Cutters.—For gears most cutters now have a centre line round the edge which facilitates their setting. It is usual in spur wheels up to about 4 or 5 pitch to rough and finish at one traverse. In large teeth a stocking cutter, Fig. 259, generally removes the bulk of material, the finishing cut being taken by another, Fig. 260. It

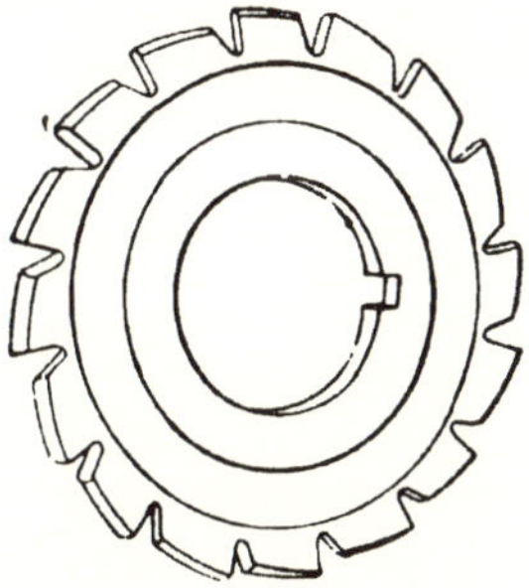

Fig. 259.—Stocking Cutter.

Fig. 260.—Gear Cutter.

is always judicious to test the first tooth and tooth space cut as to thickness before going round the wheel, and particularly so when the cutters or the machine have had much service.

The shapes of cutters are determined on a definite basis, which is that of a generating circle in the case of cycloidal teeth, that of the angle of the path contact or of pressure in involutes. In standard gears the latter is 75° with the line of centres, or $14\frac{1}{2}$° with a line cutting the line of centres at right angles. This is the equivalent of the generating circles of the cycloidal-toothed wheels. The base of the system is a rack having teeth with straight sides, each inclined at an angle of $14\frac{1}{2}$°, and therefore including an angle of 29° between them, Figs. 250 and 251. This is an extremely simple and convenient system, both in design and in cutter formation, but with wheels having less than thirty teeth it leads to some undercutting. If it is necessary to have small pinions without undercutting, then the angle of obliquity of the path of contact has to be increased to 20° or more, which means a special set of cutters designed on that basis, and a correspondingly high angle of pressure.

Manufacturers follow the Brown & Sharpe initiative in having eight cutters for a set of involute teeth of one pitch, which are well known by their numbers. As the numbers of teeth increase, the cutters include a wider range on which they will operate. A different set of cutters is properly used for spurs and bevels, the latter being thinner by 0·005 than the tooth space at the narrow end, taking the tooth length as not longer than one-third the distance from the larger diameter of the teeth to the apex of the

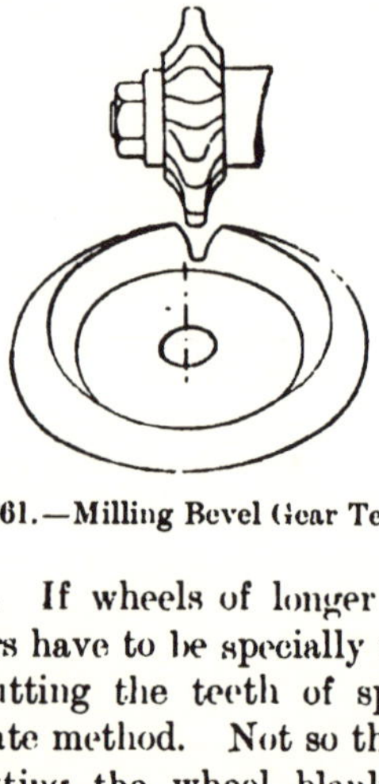

Fig. 261.—Milling Bevel Gear Teeth.

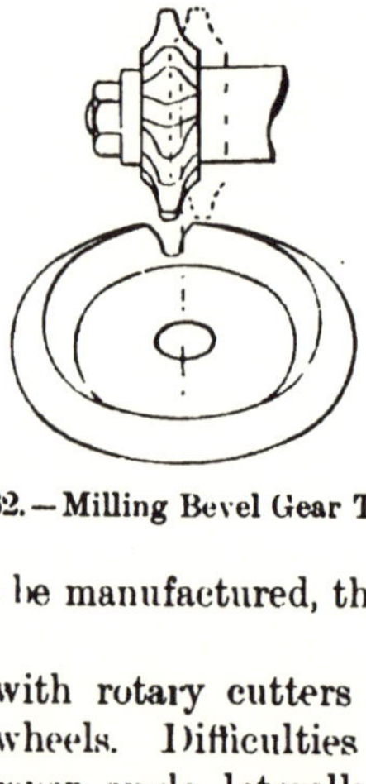

Fig. 262.—Milling Bevel Gear Teeth.

cones. If wheels of longer face have to be manufactured, thinner cutters have to be specially made.

Cutting the teeth of spur wheels with rotary cutters is an accurate method. Not so that of bevel wheels. Difficulties occur in setting the wheel blanks to the proper angle laterally, but chiefly to the fact that a given cutter can only be of correct form for one section of the wheel tooth. If we look at the section of the bevel wheel in Fig. 257, we see that the longer the teeth of a wheel are, the more unfavourable are they for fairly correct cutting; that in a length of *d* the difference in the section of the two ends is not so great as it would be in a length of *c*.

The forms of the cutters are variously taken as correct at the large end of the teeth, which is the most general practice, at a distance of one-third of the length of the teeth inwards, and, in the case of exceptionally wide teeth, at half-way along between the large and small ends. The intention, therefore, is to have the teeth true, or very approximately correct at the large end, and to allow such inaccuracies as are inseparable from the use of a single cutter to accumulate towards the smaller end. Here they will not be so very pronounced, by reason of the diminished length of teeth, and consequently of the diminished profile of the cutter in actual operation, and there will also be only the minimum of metal to be removed by the file for the purpose of correction. It is an excellent illustration of a method which is theoretically incorrect being made to produce in skilful hands fairly good results.

An important point in cutting is that the care of the profiles at the smaller ends counts for much less than shaping the widths of the tooth spaces and tooth thicknesses correctly at the pitch lines. The first may be corrected, the second not at all, or not without a good deal of trouble.

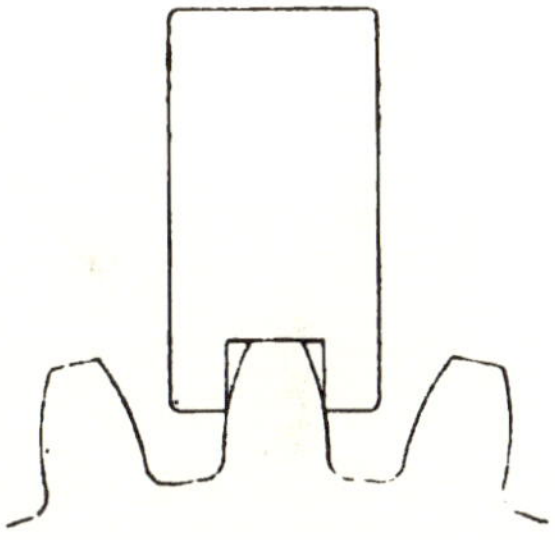

Fig. 263.—Gauging Gear Teeth.

On commencing to cut the teeth of bevel wheels, the first care is to set the blanks to the correct vertical angle, which corresponds with the angle at the bottom of the clearance space *c* in Fig. 257, keeping this parallel with the addendum length. Then the angle in the other plane is set with as close an approximation to accuracy as is practicable, Fig. 261, and a cut taken down one side. Here the need of having the cutter thinner than the tooth space at the small end of the wheel in a wide one is apparent, as a wide cutter would take out too much metal there. The gear blank is then set over at the opposite angle, and a cut taken down the opposite side of the space, Fig. 262. The widths of the space on major and minor pitch line are now checked before going further with a gauge, Fig. 263.

Projection of Bevel Wheels.—The cutters selected for mitre and for bevel wheels are not those which correspond with the numbers of teeth in the wheels, as they are in the case of spurs, but those which correspond with the number of teeth of the diameters *projected from the tooth ends* to cut the axes of the wheels. In some wheels with a slight bevel the difference will not be great; in those of flat bevel it will be very different. Only in mitre wheels, and wheels which nearly approach them, will one cutter suffice for both wheels. In most pairs of bevels a distinct cutter will be used for each of the pair.

The method of ascertaining what cutter should be selected for bevel-wheel teeth is similar to that employed for striking out the teeth on the projected faces.

In Fig. 257 the pitch diameter A of the wheel, and B of the pinion, and the actual number of teeth, are not those upon which the size of cutters is based; but the diameters A′ and B′ projected in line with the ends of the teeth to cut the centre lines have to be taken. Thus in Fig. 257 the distances *f-g*, *f-h* correspond, so far as the selection of cutters is concerned, with the radii of spur wheels. So that these distances have simply to be multiplied by 2 to obtain diameter, and the product multiplied by the diametral pitch, which gives the number of teeth that would correspond with spur wheels of radii *f-g*, *f-h*. If the radius *f-g* is 14 inches, then twice 14 inches is 28 inches, and 28 inches multiplied by a diametrical pitch, say, of $5 = 140$, the number of teeth which corresponds with the size of a wheel projected on the plane *f-g*, for which a No. 1 or "rack" cutter must be selected. For *f-h* the radius is $3\frac{1}{4}$ inches, and this multiplied by 2 gives $6\frac{1}{2}$ inches, and by the diametrical pitch $5 = 31{\cdot}25$ inches, which requires a No. 4 cutter.

This takes the cutters as being right for the large ends of the teeth, which is often done: but see page 283, except in the special cases of very long teeth. In mitre wheels, since the lengths of the projected radii would be equal, a single cutter serves for the teeth of both wheels.

The formation of the teeth by rotary cutters is done under the most favourable circumstances when the teeth are short. The proportion of one-third is given as an ordinary working limit; but this is too great in many instances. The most favourable conditions are those in which the pitch is coarse and the teeth not longer than about two and a half times the pitch. In teeth of fine pitch and of greater width, accuracy depends very much upon the care with which the cutting is supplemented by filing. In some wheels used for gearing which is required to run with exceptional smoothness at high speeds, the cutting has been supplemented by grinding in a suitable framework with emery and oil.

It is clear now why the mere possession of accurate cutters does not ensure correct bevel-wheel teeth. The cutter which is right for one end is not right for the other or for any intermediate position. Here the skill of the workman comes in. Different men adopt varied methods to get the best approximation to truth, of two evils choosing the lesser. The evil is that of having to file the faces of the teeth at the small end off to nothing at the large, as in Fig. 264, where the metal left for filing is indicated by dotted lines on the tooth on the left hand. The amount of filing can be lessened by using a cutter that gives more rounding of the faces, but then that imparts too much rounding to the faces at the large ends. Or an extra cut can be taken upon the small ends, which means another set of cuts all round on each face. Inaccuracies, too, arise due to the rolling of the gear sideways for obtaining the taper of the teeth, which tends in small pinions to cut too much off the faces, though scarcely apparent in flat bevels. The setting of the gear blank out of centre for cutting the faces to the correct bevel is not a simple matter for calculation, because the same part of the cutter which cuts the large end of the tooth at the pitch line does not cut the corresponding portion at the small end, but a narrower portion of the cutter. This is one reason why trial cuts between teeth are made at the commencement before going round, and the thicknesses of the teeth gauged at pitch lines. When the first tooth and two spaces are all right, the rest of the

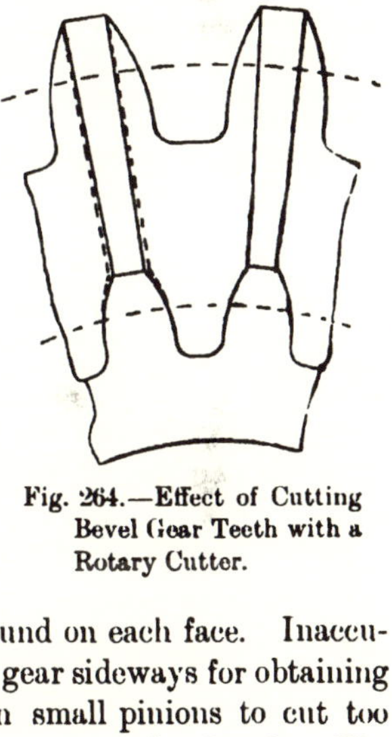

Fig. 264.—Effect of Cutting Bevel Gear Teeth with a Rotary Cutter.

cutting can be done all round one side first, followed by all the cutting on the other side.

Cycloidal as well as involute-shaped cutters are supplied for spur gears, but not for bevels, for which they would be unsuitable. If cycloidal bevels are wanted, they would have to be produced by a planing process. Comparatively few cycloidal cutters are used even for spurs, because a larger set is wanted—twenty-four for each pitch in place of eight for involutes—and because, generally,

Fig. 265.--Cutting Twelve Wheels at once.

the system is not so elastic as that of the involute. In the latter it is well known that slight differences in centres do not affect the working of the gears. A slight difference in the depth of the cutters, increasing or lessening the bottom clearance, makes no difference in the action of the gears. But in cycloidal teeth the pitch lines must roll together, and, in fact, the depth of tooth is so important that the cutters are made with shoulders which prevent cutting going below the exact depth suitable for the gear. This involves the necessity for very accurate sizing of the blanks. On the other hand, many, including the writer, think that, for smooth and easy running, double-curve cut teeth are preferable to involute, and that when the question is one of the best possible results they should be selected.

Both circular and diametrical pitches are employed in the cycloidal system of cutters, and the diameter of blanks is sized in the latter as in involutes by adding two diameter pitches to the pitch diameter.

The difference between the range of the involute and cycloidal cutters is very wide. In the first-named, three cutters, 6, 7, and 8, cut numbers of teeth from 12 to 20. But in the second, nine cutters are required, one to each separate wheel. In the higher numbers the difference is also nearly as great—three cutters from 21 to 26 teeth in the double-curve system, against one cutter from 21 to 25 in the involute, and so on.

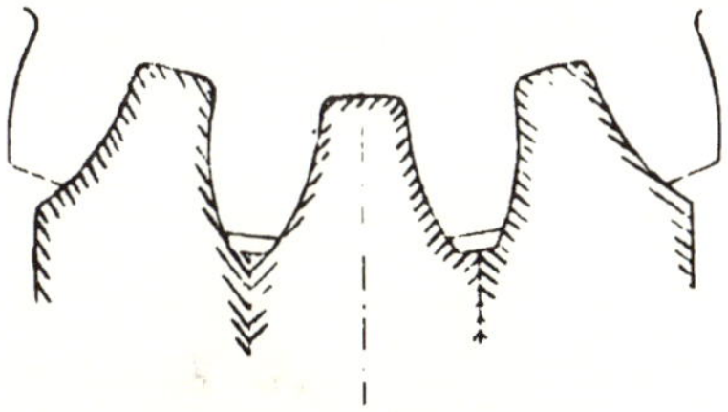

Fig. 266.—Multiple Cutters.

The gear-cutting machines have appropriated a large volume of spur and bevel gear cutting formerly done on the milling machine. Still, the work is that of rotary cutters. But the regular machines have the advantage in point of economy, being stiffer, and often designed for operating, in the case of spurs, on several spurs at once. A fine example is given in Fig. 265, where twelve wheels are being cut at once in the shops of Ludwig Loewe & Co.

The Gould & Eberhardt system of multiple cutters is another device to increase the output of a machine. Fig. 266 shows three such cutters in operation. They are, of course, only suitable for wheels of one size. On the milling machine duplication is often effected by having two or more sets of centres in which pitching is done by division plates without indexing. These are used for small gears, taps, reamers, and articles having low numbers. Racks are cut with special attachments, one or several teeth being cut at once.

Fig. 267 illustrates multiple centres by the Garvin Company

suitable for fluting taps, reamers, milling small keyways, &c. A number of small pieces are operated upon at one time, and all are indexed together, so saving a great deal of time. The spindles are geared together and are turned by a handle on the dial or by a ratchet handle. The No. 1 headstock spindles shown with $1\frac{1}{4}$-inch centres of spindles are fitted with hardened centres of No. 1 Morse taper, and these centres are cupped out and provided with four grooves, which hold the work by the square end and drive it round, so that no chucks or dogs are necessary.

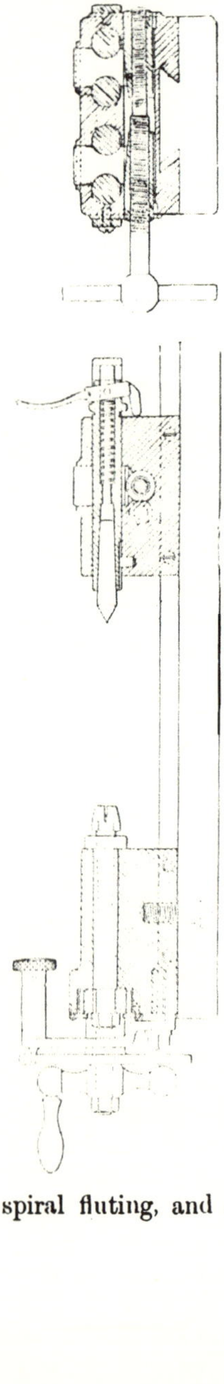

Fig. 267.—Multiple Centres.

The No. 2 headstock spindles have 2-inch centres of spindles, which have No. 3 Morse taper holes and are fitted with a double set of hardened centres. One set is cupped similar to the No. 1, and the other set are regular centres, each carrying a dogging jaw. These dogging centres enable ordinary round work to be handled.

The tailstocks are adjustable along the base block, and can be fastened at any point. The tail spindles have a spring movement, and are thrown back by separate levers, and are all bound simultaneously by one handle. Regular centres with Morse taper are fitted. The cutters can work close to the centre line.

Multiple-spindle centres are made with from two to six spindles, both large and small, especially adapted for certain purposes, such as fluting taps and reamers, making milling cutters, hardware specialities, &c. They are built of several centre distances, some being arranged for spiral fluting, and others being made to adjust for taper work.

They are designed to index simultaneously. These centres are for the most part set in large cast-iron pans, and arranged for pump connection to oil reservoir. Their efficiency is due to the fact that so much more work can be turned out by the same machine and the same attendance as compared with single-spindle centres.

Fig. 268 shows an indexing device in the Ludwig Loewe shops for milling squares on two reamers or two taps at one time.

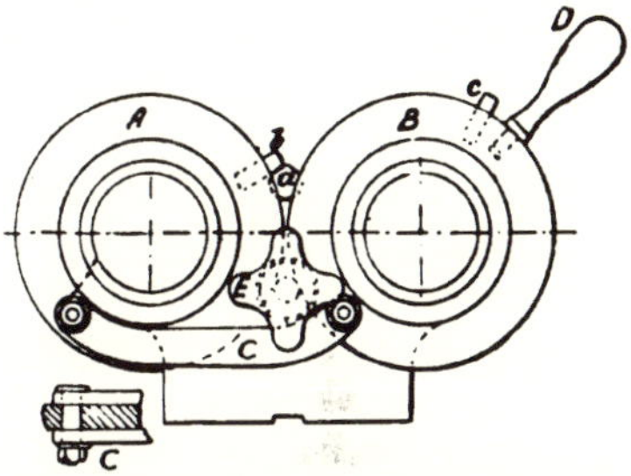

Fig. 268.—Indexing Device for Squares.

The two discs A, B are on the indexing spindles, and are connected by a link C, so that when moved by the handle D they must turn in unison. Their movement is arrested in one direction by the index pin *b* coming against the stud *a*, and in the direction opposite by the pin *c* coming on *a*, which difference is exactly equal to one-fourth of a turn. The clamp E pinches the discs when set, and the square necks are milled by straddle mills. Two mills thus cut four opposite faces at each setting.

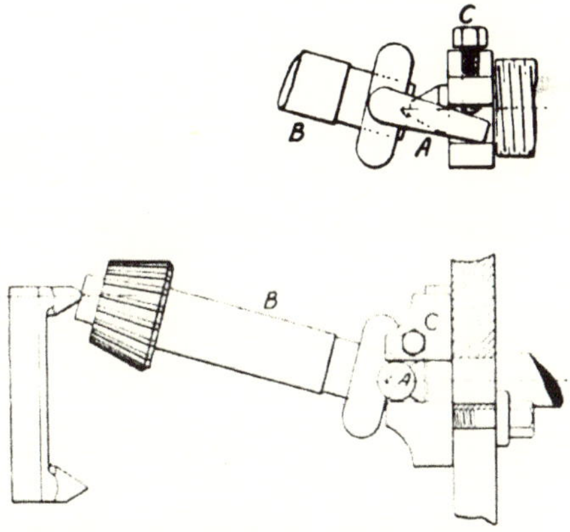

Fig. 269.—Improved Dog for Tapered Work.

The upper figure in Fig. 269 is intended to show the effect of doing tapered work between centres on the milling machine. A is the bent tail of the dog encircling the work B, and pinched by the set screw C. The result is that the tail is always wriggling about angularly and longitudinally relatively to the set screw, springing and straining the work.

The lower figure is designed to obviate this. The tail A is of cylindrical form, and is off-set so that its axis is in line with the end of the work B when the dog is set flush. An adjustable clamp C is attached to the regular driver, which holds any size of dog, and permits the tail to swivel and slide.

TABLE TO FACILITATE CALCULATIONS INVOLVING FRACTIONS.

$\frac{1}{64}$ = ·01563	$\frac{17}{64}$ = ·26563	$\frac{33}{64}$ = ·51563	$\frac{49}{64}$ = ·76563
$\frac{1}{32}$ = ·03125	$\frac{9}{32}$ = ·28125	$\frac{17}{32}$ = ·53125	$\frac{25}{32}$ = ·78125
$\frac{3}{64}$ = ·04688	$\frac{19}{64}$ = ·29688	$\frac{35}{64}$ = ·54688	$\frac{51}{64}$ = ·79688
1-16 = ·0625	5-16 = ·3125	9-16 = ·5625	13-16 = ·8125
$\frac{5}{64}$ = ·07813	$\frac{21}{64}$ = ·32813	$\frac{37}{64}$ = ·57813	$\frac{53}{64}$ = ·82813
$\frac{3}{32}$ = ·09375	$\frac{11}{32}$ = ·34375	$\frac{19}{32}$ = ·59375	$\frac{27}{32}$ = ·84375
$\frac{7}{64}$ = ·10938	$\frac{23}{64}$ = ·35938	$\frac{39}{64}$ = ·60938	$\frac{55}{64}$ = ·85938
1-8 = ·125	3-8 = ·375	5-8 = ·625	7-8 = ·875
$\frac{8}{64}$ = ·14063	$\frac{25}{64}$ = ·39063	$\frac{41}{64}$ = ·64063	$\frac{57}{64}$ = ·89063
$\frac{5}{32}$ = ·15625	$\frac{13}{32}$ = ·40625	$\frac{21}{32}$ = ·65625	$\frac{29}{32}$ = ·90625
$\frac{11}{64}$ = ·17188	$\frac{27}{64}$ = ·42188	$\frac{43}{64}$ = ·67188	$\frac{59}{64}$ = ·93188
3-16 = ·1875	7-16 = ·4375	11-16 = ·6875	15-16 = ·9375
$\frac{13}{64}$ = ·20313	$\frac{29}{64}$ = ·45313	$\frac{45}{64}$ = ·70313	$\frac{61}{64}$ = ·95313
$\frac{7}{32}$ = ·21875	$\frac{15}{32}$ = ·46875	$\frac{23}{32}$ = ·71875	$\frac{31}{32}$ = ·96875
$\frac{15}{64}$ = ·23438	$\frac{31}{64}$ = ·48438	$\frac{47}{64}$ = ·73438	$\frac{63}{64}$ = ·98438
1-4 = ·25	1-2 = ·5	3-4 = ·75	1 = 1·00000

CHAPTER XI.

FEEDS AND SPEEDS.

Governing Conditions—Feeds of more Importance than Speeds—Hardness and Softness of Metal—Pickling—Its Limitations—Frequency of Grinding—Examples of Feeds and Speeds.

Governing Conditions.—Feeds and speeds are dependent on several circumstances, such as hardness or softness of metal, the character of the cutters, whether finely or coarsely pitched; on the stiffness of the machine and arbor, or whether castings and forgings have been pickled or not, besides other conditions of a minor character. The frequency of grinding is an important item also. If a cutter is reground frequently, it will stand much more than one which is worked for several days in a more or less dull condition. Then it becomes a question of the greater economy of slowing down to suit a dull cutter, or stopping for a while to change and grind the cutter.

In making comparisons of speeds and feeds account must be taken of all these. And a distinction fully as important as any is the difference between ordinary work, and record work made as tests. It is easy by using sharp cutters, running for short periods, to make a big show by comparison with work done under ordinary shop conditions. The question then is an economical one. Record work as a rule is not paying work, and therefore not much is done in that way in ordinary shop practice. These matters, which are discussed at greater length in different sections of this work, have to be all well weighed in making comparisons between milling speeds and feeds.

Speaking broadly, the same rule obtains as in the work of single-edged cutting tools. That under identical conditions a slow cutting speed and deep cutting may be combined, while a high speed is only possible with shallow cutting.

Then, further, the rate of feed must depend largely on the degree of accuracy or finish required. No accurate work can be done on light articles with heavy cutting. The lighter the articles and the greater the amount of accuracy needed, the lighter must be the depth of cut, and rate of feed. In all work, even where the roughing cuts are heavy, the finishing cuts must be light—that is, the depth of cut must be small, though the speed and feed may be increased beyond those used for the roughing.

The feeds of milling cutters are of more importance than the speeds. Feeds should be made as fast as the cutters will stand, consistently with good work. Average surface speeds are usually given as 20 feet per minute for steel, 40 for cast iron, and 60 for brass. The speeds in gear cutting are higher, because the machines are designed specially and stiffly for that work alone: 40 feet per minute for steel, 60 for cast iron, and 80 for brass represent average speeds. The speeds of high-speed cutters can be increased by from 50 to 75 per cent. above those of ordinary steel.

TABLE OF AVERAGE CUTTING SPEEDS, PERIPHERY SPEED OF CUTTER (IN FEET) PER MINUTE.

	Brass.	Wrought Iron.	Cast Iron.	Cast Steel.
Roughing .	80	40	30	20
Finishing .	100	60	40	25
Feed per minute	2½ in.	¾ in. to 2 in.	½ in. to 1½ in.	¼ in. to ½ in.

The *rate of feed* will vary from $\frac{1}{80}$ inch to $\frac{1}{16}$ inch per foot of cutter speed, according to the strength of the work or the cutter and arbor, or the finish required and the material operated on. Thus, taking a 4-inch diameter of cutter, and maintaining the average cutting speed of 41 feet per minute (say 40 revolutions of cutter), the rate of feed may require to vary from ½ inch to 2½ inches per minute, according to the various conditions of the work, and also the breadth of cut; as a rule, the broader the cutter and the deeper the cut the slower should be the feed, although this to a great extent depends upon the power and stability of the machine.

TABLE OF FEEDS IN RELATION TO DEPTH AND WIDTH OF CUT IN ORDINARY CAST IRON, GIVING REVOLUTIONS OF CUTTER FOR 1 INCH OF FEED.

Diameter of cutters	2 in.		3 in.		4 in.		6 in.	
Width of cut	½ in.	2 in.	½ in.	2 in.	½ in.	2 in.	½ in.	2 in.
Revolutions, roughing ⅛ in. to ½ in. deep	45	40	30	25	25	22	15	12
Finishing, $\frac{1}{64}$ in. deep	35	30	22	20	18	15	10	9

This will average for roughing $\frac{1}{20}$ inch
This will average for finishing $\frac{1}{16}$ inch
{ feed of work to each 12 inches of cutter circumference, or each 4 inches of cutter diameter.

For broader cuts and harder material the number of revolutions per inch of feed should be increased.

Hardness or softness of metal makes a great difference in the results obtained by the use of milling cutters. These differences will easily halve, or double the feeds and speeds, and weight of metal removed. In any comparisons, therefore, of this kind the quality of the metal or alloy must be known. Iron castings vary greatly in hardness, covering a wide range of differences. Certain grades of steel are very tough, and must be annealed before tooling, and then they and the cutters require a large volume of lubricant.

Cast metals may be soft, homogeneous, and cut sweetly; or they may be hard, honeycombed with blow-holes, interspersed with cold shuts or chilled parts, which strain or break the tools. Forged metal may be clean and homogeneous, or it may be spilly, seamy, open, trying to the tools.

Pickling has its limitations. It cannot be conveniently practised in the case of big work. Neither is it worth while in smaller pieces, portions only of which have to be tooled, as, say, valve faces, or cover plates, or bosses. Neither is it of much utility in work where the allowances for tooling are considerable, as in many hand-made forgings, or those such as cranks, in which webs are milled from the solid, or in castings where allowances of ¼ inch or ⅜ of an inch are made for facing-off feet or flanges. There are ready means of getting below the skin in such cases, either by single-edged planer tools, or by rotary cutters with inserted teeth. But where pickling scores is in small forgings and castings where allowances are cut uniformly fine, say from $\frac{1}{32}$ inch to $\frac{1}{16}$ inch, as in stamped work, and in machine-moulded work, and in which the shallowness of the allowance would not permit the cutters to get below the hard skin and scale.

The frequency of grinding cutters has a very close relation to the volume of work which can be got through in a given time. Frequent grinding means faster and deeper cutting, but it also involves frequent stoppage of the machine for taking out and replacing cutters. If cutters are kept in duplicate, the loss of time is not so great as when the machine has to stand while regrinding is being done. For doing record work it is understood that cutters must be kept very sharp. But it is none the less desirable, in doing the ordinary work of the shop only, that a reasonable mean must be struck. Too frequent stoppages mean loss, sharpening too long delayed involves losses of time also, but in other ways. These losses are due to the necessity for slowing down or feeding more finely, and to the inefficiency of the cutter to produce true results, due to spring caused by the friction of its dulled edges. Of the two it is better to err on the side of frequent grinding than in the other way.

A list of speeds and feeds is given below, taken from various sources. Its value lies in the guidance it affords in practice. General statements are nearly valueless, but these figures give rates under all conceivable conditions, for all classes of cutters, on all kinds of machines, and they should therefore be helpful for reference and check.

The thickness of a chip can be ascertained by dividing the table feed per revolution of cutter by the number of teeth. If the feed were 0·300 inch per revolution, and the teeth 30 in number, the thickness of each chip would be

$$\frac{0·300}{30} = 0·010,$$

or a hundredth of an inch thick. This is extremely thin by comparison with the chip removed by a single-edged tool, yet it is fairly coarse for a milling chip.

The following is from the shops of A. Herbert Ltd., on fine classes of work, and in the ordinary course of practice without any attempt to establish records, some of which are shown in previous photos.

A gang of cutters milling the faces of hexagon nuts of mild steel held in a turret head, six faces on three nuts being done simultaneously. Cutters of ordinary tool steel, 6 inches in diameter, 36 teeth, 46 revolutions per minute, equal to 72 feet per minute, feed $\frac{1}{2}$ inch per minute. This gives a cut per tooth of 0·0003 inch. The heads of thirty-five $\frac{5}{8}$-inch bolts were milled per hour.

A form mill tooling the sides of links of mild steel for chain conveyors; diameter of largest part of cutter 4 inches. The total width of cut $9\frac{5}{8}$ inches, the number of teeth 14, 48 revolutions per minute, equal to 50 feet per minute, feed $\frac{1}{2}$ inch per minute, giving a feed per tooth of 0·0007 inch.

A gang of saws cutting slits $\frac{1}{32}$ inch wide in gas burners of cast iron, cutting 120 slits in one batch with 120 saws. Saws 3 inches in diameter, $\frac{1}{32}$ inch wide, number of teeth 30, 51 revolutions per minute, equal to 40 feet per minute, feed 2 inches per minute, or feed per tooth ·0013·

A form mill profiling bars of mild steel. Cutter 5 inches diameter, by $3\frac{3}{4}$ inches wide, with 12 teeth. Speed 32 revolutions per minute, equal to 42 feet per minute, feed $\frac{1}{2}$ inch per minute, giving a feed per tooth of 0·0013 inch.

A gang of three saws cutting off blanks for screwing dies of ordinary tool steel. Cutters 4 inches diameter, by $\frac{1}{4}$ inch wide, 36 teeth, 32 revolutions per minute, equal to 33·5 feet per minute, feed $\frac{1}{2}$ inch per minute, giving a feed per tooth of 0·00043 inch.

A face mill with inserted cutters of Armstrong-Whitworth high-speed steel, and having front rake, surfacing capstan slides. The cutter was 12 inches in diameter, with 16 teeth, $11\frac{1}{2}$ inches width of cut, $\frac{5}{32}$ inch depth of cut, 24 revolutions per minute, equal to 75 feet per minute, feed $7\frac{1}{2}$ inches per minute, or a feed per tooth of 0·015 inch.

The foregoing, the last excepted, were done on horizontal spindle machines of the pillar and knee type, the arbor being in each of the examples supported at the outer end.

The following examples are of work done with rotary face mills, with inserted cutters, on horizontal spindles.

Tooling castings very hard and with hard spots, and not pickled. Cutter 27 inches diameter, 36 teeth, width of cut

15 inches, depth $\frac{1}{8}$ inch to $\frac{3}{16}$ inch, speed $2\frac{1}{2}$ revolutions per minute, equal to 17 feet cutting speed, feed $3\frac{1}{2}$ inches per minute, giving a feed per tooth of 0·038 inch. The cutters were of Mushet steel, and tempered once only in 10 days.

Tooling castings of similar hard quality. Cutter 40 inches diameter, with 52 teeth, width of cut 32 inches, depth $\frac{3}{16}$ inch to $\frac{5}{16}$ inch in places, speed $1\frac{2}{3}$ revolutions per minute, equal to 17 feet cutting speed, feed 3 inches per minute, giving a feed per tooth of 0·033 inch. The cutters were made of Mushet steel, sharpened once in from twelve to fifteen days.

Castings not so hard, but rough. Cutter 11 inches in diameter, with 8 teeth, width of cut 8 inches, depth $\frac{1}{8}$ inch, 7 revolutions per minute, equal to 20 feet cutting speed, feed 2 inches per minute, giving a feed per tooth of 0·033 inch. Cutters made of Armstrong-Whitworth steel, sharpened once in four days.

An inserted toothed slabbing cutter operating on a plane surface of cast iron, took a cut $8\frac{1}{4}$ inches wide, with a depth ranging from $\frac{3}{32}$ inch to $\frac{5}{32}$ inch from a scale surface, at a speed of cutter of 40 feet per minute, and a rate of feed of $8\frac{1}{2}$ inches per minute.

Examples of milling a tee groove in milling-machine platens. In one the body of the slot was first roughed out, and the bottom done on the milling machine, with a rate of feed of $6\frac{1}{2}$ inches per minute. In another the tee slot was roughed out with an end mill, at a feed of $6\frac{1}{2}$ inches per minute, after which the tee was done at a feed of $6\frac{3}{8}$ inches per minute. This was only possible by the application of a compressed air blast to clear out the chips.

The following are from the shops of the Cincinnati Milling Machine Company, using Novo steel.

In milling gibs of grey cast iron, 2 inches wide, an edge cutter 4 inches diameter, with 18 teeth, which it will be noted is a coarse pitch, is driven at a peripheral speed of 90 feet per minute, with a feed of 0·300 inch per revolution of the cutters. The strip travels under the cutter at a rate of 27 inches per minute; and the cut is $\frac{3}{32}$ inch deep.

The finishing cut is taken by a face mill, removing 0·010 inch in depth, with a feed of 0·200 inch per revolution of cutter, or 10 inches per minute.

Another is milling cover plates of grey iron, measuring

6×4 inches. A slabbing spiral-toothed cutter, 3 inches diameter, with 16 teeth, makes 51 revolutions per minute, feeds at 0·210 inch per revolution, or 10·7 inches per minute, and removes $\frac{1}{8}$ inch depth of metal.

Another job is milling tee slots in the tables of the firm's milling machines. These are first milled over the face, and the vertical cut for the slots taken. Then a stem cutter, measuring $1\frac{1}{16}$ inch $\times$ $\frac{1}{2}$ inch, with 8 teeth, tee-slots each groove at a travel of 12 inches per minute. The chips are removed by an air blast.

In these shops, extra duty is regularly got out of the milling cutters by utilising a blast of compressed air to drive the chips away from the cutters. Branch pipes are led to the machines, and a hose having a nozzle, with a spring-closed valve, is opened by a slight pressure of the thumb. Without the air blast, the chips would clog or break the cutters in heavy horizontal milling.

Another device employed is that of a strong jet of oil delivered against the cutters, with the same object—that of clearing away the chips.

The Holroyd patent pressed steel milling cutters have the following records:—

Two ordinary steel patent cutters, $2\frac{3}{4}$ inches diameter, running 600 revolutions per minute, cutting drawn brass screws $\frac{1}{2}$-inch diameter, $\frac{1}{2}$-inch lead, $\frac{1}{4}$-inch pitch, width and depth of cut $\frac{1}{16}$ inch, feed $6\frac{1}{2}$ inches per minute—these were ground once each week.

Two high-speed steel cutters, cut on three edges, $2\frac{3}{4}$ inches diameter, running 200 revolutions per minute, cutting bright drawn screw 1-inch diameter, $\frac{1}{2}$-inch lead, $\frac{1}{4}$-inch pitch, width and depth of cut $\frac{1}{8}$ inch, feed 4 inches per minute—these were ground every 11 hours.

One ordinary steel cutter, cut on three edges, $3\frac{1}{4}$ inches diameter, running 90 revolutions per minute, cutting mild steel screw $3\frac{1}{2}$ inches diameter, $3\frac{1}{2}$ inches lead, 3 starts, width of cut $\frac{9}{16}$ inch, depth of cut $\frac{7}{16}$ inch, feed $1\frac{1}{4}$ inch per minute—this was ground every 5 hours.

Two high-speed steel cutters, cut on three edges, $2\frac{3}{8}$ inches diameter, running 200 revolutions per minute, cutting bright drawn screw $\frac{3}{8}$-inch diameter, $\frac{1}{2}$-inch lead, $\frac{1}{4}$-inch pitch, width and depth of cut $\frac{1}{16}$ inch, feed $4\frac{1}{2}$ inches per minute—these were ground every 9 hours.

One high-speed steel cutter, cut on three edges, $2\frac{3}{4}$ inches diameter, running 1,700 revolutions per minute, cutting cast brass screw 1-inch diameter, $\frac{1}{4}$-inch pitch, $\frac{1}{4}$-inch lead, width and depth of cut $\frac{1}{8}$ inch, feed 60 inches per minute—there was no appreciable wear after the cutter had milled a gross of these screws, each 2 inches long.

Also two high-speed armour-plåte bolt cutters made of Armstrong-Whitworth steel—one of the cutters, making 160 revolutions per minute, cutting 6 inches peripheral speed, was ground every 8 hours; the other, making 200 revolutions per minute, and cutting 12 inches per minute feed, lasted 2 hours.

A pair of high-speed steel cutters $2\frac{3}{4}$ inches diameter, worked eight weeks continuously, and cut 3,000 feet of double thread. When new they were $2\frac{3}{8}$ inches diameter, and lost $\frac{1}{8}$ inch diameter through grinding in eight weeks. The cutters were still in fair condition.

MILLING CUTTERS.

Table of Cutting Speeds. (See opposite page.)

(Morse Twist Drill & Machine Company.)

Tho table will be convenient for finding the number of revolutions per minute required to give a periphery speed from 5 feet to 50 feet per minute of diameters from $\frac{1}{2}$ inch to 30 inches.

Examples.—A mill 2 inches in diameter, to have a periphery speed of 35 feet per minute, should make about 67 revolutions, while a $1\frac{1}{8}$-inch mill should make 120 revolutions to have the same periphery speed. If a $\frac{3}{4}$-inch mill makes 250 revolutions per minute, the periphery speed is about 50 feet.

Table of Cutting Speeds.

Feet per Minute	5	10	15	20	25	30	35	40	45	50
Diam	Revolutions Per Minute									
Inches.										
½	38·2	76·4	114·6	152·9	191·1	229·3	267·5	305·7	344·0	382·2
⅝	30 6	61·2	91·8	122·5	153·1	183·7	214·3	244·9	275·5	306·1
¾	25·4	50·8	76·3	101·7	127·1	152·5	178·0	203·4	228·8	254·2
⅞	21·8	43·6	65·5	87·3	109·1	130·9	152·7	174·5	196·3	218·9
1	19·1	38·2	57·3	76·4	95·5	114·6	133·8	152·9	172·0	191·1
1⅛	17·0	34·0	51·0	68·0	85·0	102·0	119·0	136·0	153·0	170·0
1¼	15·3	30·6	45·8	61·2	76·3	91·8	106·9	122·5	137·4	153·1
1⅜	13·9	27·8	41·7	55·6	69·5	83·3	97·2	111·1	125·0	138·9
1½	12·7	25·4	38·2	50·8	63·7	76·3	89·2	101·7	114·6	127·1
1⅝	11·8	23·5	35·0	47·0	58·8	70·5	82·2	93·9	105·7	117·4
1¾	10·9	21·8	32·7	43·6	54·5	65·5	76·4	87·3	98·2	109·1
1⅞	10·2	20·4	30·6	40·7	50·9	61·1	71·3	81·5	91·9	101·9
2	9·6	19·1	28·7	38·2	47·8	57·3	66·9	76·4	86·0	95·5
2¼	8·5	17·0	25·4	34·0	42·4	51·0	59·4	68·0	76·2	85·0
2½	7·6	15·3	22·9	30·6	38·2	45·8	53·5	61·2	68·8	76·3
2¾	6·9	13·9	20·8	27·8	34·7	41·7	48·6	55·6	62·5	69·5
3	6·4	12·7	19·1	25·5	31·8	38·2	44·6	51·0	57·3	63·7
3½	5·5	10·9	16·4	21·8	27·3	32·7	38·2	43·6	49·1	54·5
4	4·8	9·6	14·3	19·1	23·9	28·7	33·4	38·2	43·0	47·8
4½	4·2	8·5	12·7	16·9	21·2	25·4	29·6	34·0	38·1	42·4
5	3·8	7·6	11·5	15·3	19·1	22·9	26·7	30·6	34·4	38·2
5½	3·5	6·9	10·4	13·9	17·4	20·8	24·3	27·8	31·3	34·7
6	3·2	6·4	9·6	12·7	15·9	19·1	22·3	25·5	28·7	31·8
7	2·7	5·5	8·1	10·9	13·6	16·4	19·1	21·8	24·6	27·3
8	2·4	4·8	7·2	9·6	11·9	14·3	16·7	19·1	21·1	23·9
9	2·1	4·2	6·4	8·5	10·6	12·7	14·9	17·0	19·1	21·2
10	1·9	3·8	5·7	7·6	9·6	11·5	13·4	15·3	17·2	19·1
11	1·7	3·5	5·2	6·9	8·7	10·4	12·2	13·9	15·6	17·4
12	1·6	3·2	4·8	6·4	8·0	9·6	11·1	12·7	14·3	15·9
13	1·5	2·9	4·4	5·9	7·3	8·8	10·3	11·8	13·2	14·7
14	1·4	2·7	4·1	5·5	6·8	8·1	9·6	10·9	12·3	13·6
15	1·3	2·5	3·8	5·1	6·4	7·6	8·9	10·2	11·5	12·7
16	1·2	2·4	3·6	4·8	6·0	7·2	8·4	9·6	10·7	11·9
17	1·1	2·2	3·4	4·5	5·6	6·7	7·9	9·0	10·1	11·2
18	1·1	2·1	3·2	4·2	5·3	6·4	7·4	8·5	9·6	10·6
19	1·0	2·0	3·0	4·0	5·0	6·0	7·0	8·0	9·1	10·1
20	1·0	1·9	2·9	3·8	4·8	5·7	6·7	7·6	8·6	9·6
21	·9	1·8	2·7	3·6	4·5	5·5	6·4	7·3	8·1	9·1
22	·9	1·7	2·6	3·5	4·3	5·2	6·1	6·9	7·8	8·7
23	·8	1·7	2·5	3·3	4·1	5·0	5·8	6·6	7·5	8·3
24	·8	1·6	2·4	3·2	4·0	4·8	5·6	6·4	7·2	8·0
25	·8	1·5	2·3	3·1	3·8	4·6	5·3	6·1	6·9	7·6
26	·7	1·5	2·2	2·9	3·7	4·4	5·1	5·9	6·6	7·3
27	·7	1·4	2·1	2·8	3·5	4·2	5·0	5·7	6·4	7·1
28	·7	1·4	2·0	2·7	3·4	4·1	4·8	5·5	6·1	6·8
29	·7	1·3	2·0	2·6	3·3	4·0	4·6	5·3	5·9	6·6
30	·6	1·3	1·9	2·5	3·2	3·8	4·5	5·1	5·7	6·4

INDEX.

Printed at THE DARIEN PRESS, *Edinburgh.*

Spotted something that isn't quite right?

Let us know and we'll fix it for future reprints. And as a thank you for your diligence, we'll send you a free Sothis book of your choice.

publisher@sothis.press

This page intentionally left blank

Made in United States
Orlando, FL
05 November 2026

90231929R00187